Noah's Ark of
the 21st Century

Animal Science, Issues and Professions

Additional books in this series can be found on Nova's website under the Series tab.

Additional E-books in this series can be found on Nova's website under the E-book tab.

NOAH'S ARK OF THE 21ST CENTURY

JOSEPH SARAGUSTY AND AMIR ARAV

Nova Science Publishers, Inc.

New York

NOTICE TO THE READER

Library of Congress Cataloging-in-Publication Data
Saragusty, Joseph.
  Noah's ark of the 21st century / authors, Joseph Saragusty, Amir Arav.
     p. cm.
  Includes index.
  ISBN 978-1-61324-492-0 (hardcover)
  1. Wildlife management. 2. Wildlife conservation. I. Arav, Amir. II. Title.
  SK355.S27 2011
  639.9--dc23
                              2011012978

Published by Nova Science Publishers, Inc. † New York

CONTENTS

PREFACE

During the course of evolution species have always gone extinct; however, the rate of extinction has increased in recent decades by as much as one thousand fold. *In situ* preservation should be supported by *ex situ* efforts like captive breeding, supplemented by assisted reproductive technologies and the establishment of genome resource banks.

Semen cryopreservation protocols have been developed for many species but many others proved challenging. Apparently, specie-specific protocols for semen collection and cryopreservation should be developed. We, and others (Jewgenow et al. 1997; Saragusty et al. 2006), have shown that post mortem semen collection and cryopreservation from endangered species, even hours after death, can save valuable genes. To reduce costs of liquid nitrogen storage and maintenance, we have developed a large volume cryopreservation technique. Additionally, with this cryopreservation technique we showed that samples could be thawed, used and the balance refrozen (Arav et al. 2002b; Saragusty et al. 2009c). Other mid- and long-term preservation techniques are currently under exploration, including freeze-drying (Loi et al. 2008a) and electrolyte-free preservation (Saito et al. 1996).

Oocyte cryopreservation has proved much more challenging than sperm and, even today, success rate is very low. Vitrification is gaining the lead in this field. However, as we have recently demonstrated (Yavin and Arav 2007), identifying the delicate balance between multiple factors is imperative for success. Alternatively, oocytes collected ante- or post mortem can be fertilized *in vitro* and cryopreserved as embryos.

Embryo cryopreservation technology, initially by slow freezing and more recently also by vitrification, has been in practice now for four decades but it is still nearly exclusively in use for laboratory animals, livestock and humans.

Relatively little progress has been achieved in wildlife species, and the little that has been done took place almost solely in mammals. Embryo cryopreservation, however, holds great promise for wildlife conservation due to its compaction of both maternal and paternal genomes into a single, transferable unit.

Ovary freezing (cortical slices or whole ovary) is a new technology developed for human fertility preservation in women that undergo cancer treatment. In a recent publication we (Arav et al. 2010) documented the longest ovarian function for up to 6 years after whole organ cryopreservation in sheep. We have shown endocrine cyclicity and production of normal oocytes and embryos. This strategy could benefit endangered species if allogeneic or even xeno-transplantation after whole ovary or ovarian tissue cryopreservation could be done.

Cloning animals is currently limited to few species and has a low success rate. Nevertheless, despite this limitation, it would seem pragmatic to initiate storage of somatic cells with an eye to future improvements in nuclear transfer efficiency. However, a major obstacle to the establishment of genome resource banks is the cost associated with the long-term maintenance of cell lines under liquid nitrogen. We have shown recently (Loi et al. 2008a; Loi et al. 2008b) the capacity of freeze-dried somatic cells, which were held at room temperature for three years to maintain their nuclear integrity, and subsequently be used for nuclear transfer and produce viable embryos.

In the following pages, we will review the various aspects of gametes, embryo and tissue preservation for prospective utilization in assisted reproductive technologies.

INTRODUCTION

The Species Survival Commission (SSC) of the International Union for Conservation of Nature and Natural Resources (IUCN) continuously monitors the planet's fauna and flora and launches the IUCN Red List of Threatened Species (http://www.iucnredlist.org). As of the end of 2010, there were 5491 species of mammals described. Of these, 1,131 species (21%) are now listed as endangered to some degree. In addition, there are 324 species listed as near threatened and another 836 species for which data is deficient and thus could be at risk. Adding all these numbers together, about 42% of the planet's mammalian species are at some level of threat for extinction. The list also reports on 76 species (1.4%) of mammals that became extinct in recent years and 2 more species that are extinct in the wild and whose survival completely depend on *ex situ* conservation programs. The situation is not distinctively different in other classes of the vertebrata subphylum or in the other subphylums of the animal kingdom. If anything, it is even worse for some such as the reptiles (21% endangered), amphibians (30%), fish (21%) or among the invertebrates: insects (22%), mollusks (41%), crustaceans (28%), anthozoa (corals and sea anemones; 27%) or arachnids (58%). With each extinct species, the stability of the entire ecological system surrounding it and the food chain of which it is an integral part is shaken. Such shaking may lead to the co-extinction of dependent species (Koh et al. 2004).

In 1992 the Convention on Biological Diversity (CBD) was ratified at the United Nations Conference on Environment and Development in Rio de Janeiro. Ten years later, during the 6th meeting of the Conference of the Parties to the CBD, it was agreed "to achieve by 2010 a significant reduction of the current rate of biodiversity loss at global, regional and national level as a contribution to poverty alleviation and to the benefit of all life on Earth" (Convention on Biological Diversity 2002). However, 2010 has arrived and this target not only has not been met, even some of the indicators needed to measure progress (or regress) have not yet been developed or fully implemented (Walpole et al. 2009). Based on paleontological data, of the total biota of about 10 million species, the natural or background extinction rate is approximately 1 to 10 species per year (Reid and Miller 1989). This may be divided into species with restricted ranges for which extinction rate might be higher and those with widespread ranges for which it is considerably lower. The expected extinction rate amongst all bird and mammal species is about one species every 100 to 1,000 years, yet the current extinction rate for these and other groups is about one species per year, which is 100 to 1,000 times the natural rate (Ceballos and Ehrlich 2002; IUCN 2004; Living Planet Report 2008; Reid and Miller 1989). Earth history has witnessed 5 major events of mass extinctions in which a significant fraction of the diversity in a wide range of taxa went extinct within relatively short period (Erwin 2001). The last, and probably the most well known episode, took place during the late Cretaceous era, approximately 65 million years ago, when the dinosaurs became extinct. The current dramatically accelerated rate of species extinction has been likened to these evens and was termed 'the sixth mass extinction event in the history of life on Earth' (Chapin et al. 2000; Wake and Vredenburg 2008) Various studies have demonstrated the severity of this mass extinction process on both the population level (Ceballos and Ehrlich 2002) and the global biodiversity level (Living Planet Report 2008; Rockstrom et al. 2009). This sixth mass extinction is anthropogenic in essence, resulting from five major human interference categories: (i) habitat loss or fragmentation, (ii) over exploitation, (iii) species introduction (exotic species and diseases), (iv) pollution of water, soil and air, and (v) global warming. Based on different projections such as climate change, human population growth or deforestation rate, predictions suggest that large chunks of the world's biodiversity is destined to disappear (Ehrlich and Wilson 1991; Reid and Miller 1989; Thomas et al. 2004).

Conservation can take one of two basic approaches. One approach is conservation of the habitat selected based on the biodiversity in it (Margules and Pressey 2000). Another is the conservation effort directed at individual species, primarily under captive conditions, through natural breeding and assisted reproductive technologies (Holt and Pickard 1999; Pukazhenthi and Wildt 2004; Wildt 1992). Unfortunately, conservation often conflicts with the rapidly growing world's human population and its ever-increasing demand for land, food, water and energy. Society will obviously elect to satisfy humans' needs before any consideration of conservation. However, ignoring the issue of conservation will be the wrong approach. Nature and biodiversity provide us with security, resiliency, social relations, health, and freedom of choices and actions as well as welfare and livelihood (Duraiappah et al. 2005). While conservation of the habitat is of paramount importance, this alone is not enough. *Ex situ* conservation efforts are necessary to complement it, and for many species these are the last chance for survival (IUCN 1987). To that end, captive breeding programs, at times supplemented by assisted reproductive technologies (ART), were set into motion for a wide variety of species. Yet, detailed knowledge of the mechanisms of reproduction is available for only about 2% of the world's mammals, many of which are domestic and laboratory animals (Comizzoli et al. 2000; Wildt et al. 1997). Regrettably, attempts to apply to wildlife species techniques developed for humans, livestock or laboratory animals, often fail to produce the expected results (Wildt et al. 1995). Such techniques are often species-specific and of little use when applied to other species. While basic evaluation techniques can usually be adapted from one species to another, species-specific methods should be devised for some or all of the following procedures: 1) semen collection and handling, 2) artificial insemination (AI), 3) oocyte collection, with or without hormonal stimulation, by ovum pick up (OPU) or post mortem, 4) *in vitro* oocyte maturation (IVM) and *in vitro* fertilization (IVF) or intracytoplasmic sperm injection (ICSI), 5) *in vitro* embryo culture (IVC) and embryo transfer (ET), timed with natural estrus or following chemical synchronization, 6) somatic cell nuclear transfer (SCNT) and 7) the preservation of gametes, embryos, tissues and whole organs (Figure 2). To store and manage the preserved cells, embryos, tissues and organs, the establishment of genome resource banks (GRB) was proposed (Holt et al. 1996; Johnston and Lacy 1995; Soulé 1991; Veprintsev and Rott 1979; Wildt 1992; Wildt et al. 1997). Apart from being a collection of genomes, gametes and embryos, fulfilling their function as a mean to extend the reproductive lifespan of individuals beyond their biological life and to prevent the loss of valuable individuals to the gene pool, these establishments provide multiple additional advantages. Transporting the preserved gametes or embryos is easier and cheaper than shipping the stress-susceptible live animal for breeding. Avoiding the transfer of animals from one institution to another can also help eliminate the risk of transferring diseases along with the transferred animal. Sperm collected from healthy individuals in captivity can be used to inseminate females in wild isolated small populations. Semen can also be collected from the wild to revitalize the captive population with new genetic material, without the need to remove valuable individuals from the wild population. Collections in GRB also act as insurance for small populations against catastrophes, epidemics etc. For example, over half of the mountain gazelle (Gazella gazella gazella) population in the north of Israel was wiped out following an outbreak of the foot and mouth disease in the mid 1980's (Shimshony 1988; Shimshony et al. 1986) or the canine distemper virus epidemic of 1994 in the lion (*Panthera leo*: Figure 1) population of the Serengeti-Mara system lead to the death of about a third of the population there (Roelke-Parker et al. 1996). In the absence of a system to collect and bank samples, all these dead animals are lost for the gene pool.

However, genetic diversity is lost only when animals are no longer available to reproduce (Ballou 1992). Gametes can be collected from animals ante mortem and stored beyond their reproductive lifespan or they can be collected post mortem. Genome resource banks may also help in mating between individuals that are incompatible due to character, personal preferences, location or time. Keeping biodiversity in GRB can also eliminate the need to keep large groups of living animals to meet targeted genetic diversity and by that reducing one of the major problems faced by many zoos around the world - space. Such collections can also be used for the purpose of research on evolving assisted reproductive technologies. Several such GRBs are already in existence. These include for example the Frozen Ark Consortium (http://www.frozenark.org/; Clarke 2009), the Amphibian Ark (http://www.amphibianark.org/), the Biological Resource Bank of Southern Africa's Wildlife (Bartels and Kotze 2006) and a number of additional institutions, some of which were listed by Andrabi and Maxwell (2007).

Long-term preservation of such biological material is almost entirely a matter of how water therein is dealt with. Seeds, whose water content is very low, can easily be preserved and at relatively high subzero temperatures of -20 to -30°C (Ruttimann 2006) whereas water content in animal tissues and cells is very high, thus requiring special handling. In the following pages difficulties and achievements concerned with fertility and genetic preservation through the preservation of gametes, embryos, tissues and organs will be reviewed.

Figure 1. Lion (*Panthera leo*) population at the Serengeti-Mara system decreased in about one third due to canine distemper virus infection. All those dead animals are lost forever to the population gene pool. Photo by Eyal Bartove ©.

Reproduction biotechnology and female genetic preservation

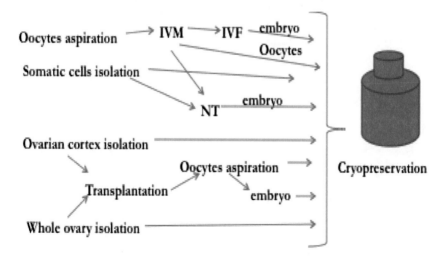

Figure 2. (Continued)

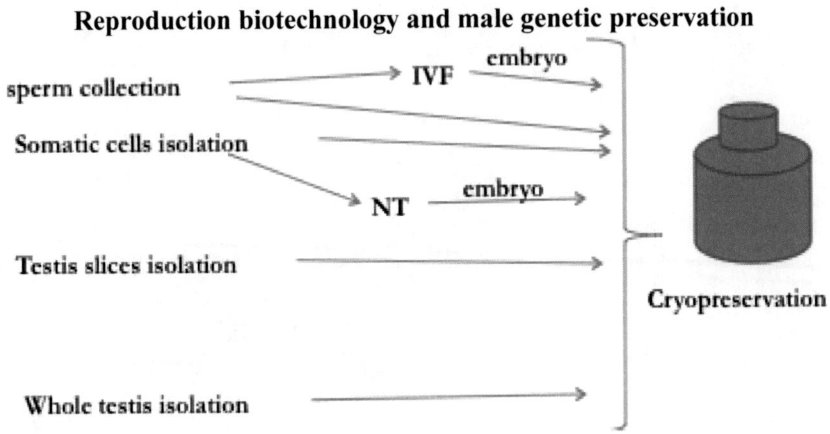

Figure 2. Reproductive biotechnology offers a variety of options for the preservation of genetic material and germplasm from both the female and the male.

Chapter 1

SPERM PRESERVATION

Probably the first step, before even starting the process of semen collection and preservation, would be to learn about the reproductive physiology of the species in question. The knowledge on similar domestic or laboratory species or even on other members of the same family is often of little help. For example, some felids such as the lynx (*Lynx lynx*) or the Pallas' cat (*Otocolobus manul*) show clear seasonality in their reproductive behavior and semen production (Göritz et al. 2006; Swanson 2006), whereas other cat species such as the South American cats show weak or no seasonality at all (Morais et al. 2002; Swanson and Brown 2004). Therefore, knowledge will indicate when would be the best time to collect semen samples. Evaluation of the reproductive tract and accessory glands through the use of ultrasonography may be of great help in this respect, both to identify pathologies and to understand how the system is built, what is the status of the accessory glands and what can be expected during collection (Hildebrandt et al. 2000a; Hildebrandt et al. 2000b). Monitoring of the reproductive steroid metabolites in feces, urine and saliva is a non-invasive method for the evaluation of the reproductive status of animals, suitable for animals both in zoos and in the wild (Brown et al. 1996; Freeman et al. 2010; Schwarzenberger et al. 1996). While this method is primarily in use for monitoring females during the reproductive cycle and pregnancy, it was also applied in recent years to males of various species (Göritz et al. 2006; Kretzschmar et al. 2004; Morais et al. 2002). Naturally, dissection of the reproductive system, when possible, can contribute valuable knowledge about the system construction, knowledge that may become helpful during semen collection and when conducting ultrasound evaluation of the reproductive tract during fertility and health examination.

After the general evaluation of the reproductive tract health and monitoring for reproductive activity, seasonality and such, collection and evaluation of semen will give additional insights into individual and species-specific reproductive characteristics.

SEMEN COLLECTION AND EVALUATION

Numerous semen collection techniques were described in the literature for a wide variety of species. Some of the techniques involve natural mating followed by collection of the semen from the female either directly from the vagina as was done, for example, in the Sumatran rhinoceros (O'Brien and Roth 2000) or through the aid of intra-vaginal condom or vaginal sponge as is done, for example, in some camelids (Bravo et al. 2000). For these techniques, one should have the right setting that will lead to natural mating and the ability to access the female immediately after. Other techniques involve extensive training of the male such as the use of artificial vagina. This method is in use in some of the equid, felid, cervid and camelid species (Asher et al. 2000; Bravo et al. 2000; Gastal et al. 1996; Zambelli and Cunto 2006) and was attempted with a lesser degree of success in other species. Other techniques that require a certain degree of training are those that involve manual stimulation of either the rectum as is routinely done in elephants (Schmitt and Hildebrandt 1998; Schmitt and Hildebrandt 2000) or through stimulation of the penis as is done for example in the flying fox (*Pteropus alecto*) or the marmoset monkey (*Callithrix jacchus*) (Melville et al. 2008; Schneiders et al. 2004). Probably by far the most popular and widely used method is the electroejaculation technique. To be successful, one would need a suitable probe, which often needs to be specifically designed for the animal to be collected based on preliminary knowledge of its anatomy (Hildebrandt et al. 2000a; Roth et al. 2005). Despite its wide use and success in many species, this technique is not without drawbacks. To begin with, the animal must be immobilized, something many zoos would rather avoid when possible. The need to anesthetize the animal makes it impossible to collect from the same individual on a regular, frequent basis. During collection, unintentional stimulation of the bladder often result in urinary contamination of the sample, as was recently reported, for example, in the Asiatic Black bears (*Ursus thibetanus*) (Chen et al. 2007). However, with proper anatomical understanding of the rectal region, adjustment of the stimulation level and, when possible, emptying the bladder prior to collection, this can be avoided.

Pharmacologically-induced ejaculation was also described in recent years. For example in stallions, oral imipramine hydrochloride followed by intravenous xylazine hydrochloride was shown to result in ejaculation minutes after the xylazine injection in 68% of the attempts (McDonnell 2001). Finally, there are the invasive techniques, which include urethral catheterization after medetomidine administration (Filliers et al. 2010; Zambelli et al. 2008), semen aspiration from the cauda epididymis for evaluation (e.g. Moghadam et al. 2005) or salvage of semen from the epididymis and proximal portion of the vas deference following castration or post mortem (e.g. Jewgenow et al. 1997; Mahesh et al. 2011; Saragusty et al. 2006; Figure 3). From our and others experience, sperm survive well as long as it is in the epididymis and chilled to $4^{\circ}C$ (Yu and Leibo 2002), even when still attached to the testicle (Saragusty et al. 2006) and live births were reported following artificial insemination, *in vitro* fertilization or intracytoplasmic sperm injection (Santiago-Moreno et al. 2006; Shefi et al. 2006).

Sperm evaluation also requires understanding of the species under study. In primates, semen can be as thick as paste, which requires liquefaction and extraction of the cells into a diluent. In camelids, possibly due to the absence of vesicular glands, sperm is also fairly viscous but it can be enzymatically liquefied (Bravo et al. 2000). It was found that similar enzymatic liquefaction can also help when attempting to separate rhinoceros sperm from the seminal plasma, something that can not be done efficiently with centrifugation alone in some of the ejaculates (Behr et al. 2009b). The volume and concentration also vary by several orders of magnitude among species. In the naked mole rat (*Heterocephalus glaber*) only 5 to 10 µL of sperm can be collected with a few hundred cells at the most (unpublished data). In the European brown hare (*Lepus europaeus*) or the Asiatic black bear for example, volume of semen collected by electroejaculation is often in the range of 1 mL or less with concentrations that often exceeding 10^9 cells/mL. Low volume of up to a few mL and low concentration of few millions per mL is often the case in felids both in captivity and in the wild (Barone et al. 1994; Morato et al. 2001). In the pygmy hippopotamus (*Choeropsis liberiensis*) the sperm-rich fraction can be extremely concentrated. In one case we found as much as 9.85×10^9 spermatozoa per mL (Saragusty et al. 2010a). In some other animals volumes can be very large. In boar, donkey or elephant semen can exceed 100 mL with concentrations of several hundred million to over a billion cells per mL (e.g. Contri et al. 2010; Saragusty et al. 2009e; unpublished data). Motility is also expected to be low in sperm collected from the epididymis, as epididymal sperm is immotile in most mammals. This is expected to change after a short incubation time in a suitable media. Some specific characteristics were also noted in certain species. For example the seminal plasma pH of the flying fox or the snow leopard (*Panthera uncial*) is high (8.2 and 8.4, respectively) (Melville et al. 2008; Roth et al. 1996) or in the Asian elephant (*Elephas maximus*) we found recently that the osmolarity of the seminal plasma was around 270 mOsm/kg (Saragusty et al. 2009e). Such characteristics demonstrate the need to verify multiple aspects of the semen so that suitable diluents can be made.

Many wildlife captive populations, such as the Asian elephant, small cat species, the Przewalski's horse (*Equus ferus przewalskii*) or the Somali wild ass (*Equus africanus somaliensis*), originated from a small founder group. Topping this with aging of the captive population, no introduction of new genetic material through new captures from the wild and no preservation of the existing genetic diversity through GRB, and the result is genetic drift that may eventually lead such populations to extinction (Moehlman 2002; Swanson 2006; Wiese 2000; unpublished data). A similar problem may arise in wild populations that become isolated due to habitat fragmentation or where the number of individuals left in the wild is small, like in Javan rhinoceros (*Rhinoceros sondaicus*) whose wild population is estimated at under 60 or the Northern white rhinoceros (*Ceratotherum simum cottoni*) whose wild population was 4 individuals at last census in 2006 and is now believed to be extinct in the wild (http://www.rhinos-irf.org/). Recently 4 out of the 8 remaining captive Northern white rhinoceroses were shipped to a reserve in Kenya, Africa with the hope that they will reproduce there, something they rarely did in captivity. More than a year has passed since and we now start seeing indications that this might be happening. Two matings were observed in January 2011, one between two Northern white rhinoceroses and one between an old Northern white rhinoceros male and a Southern white rhinoceros female (http://www.northernwhiterhinolastchance.com/news_Jan25-2011.html). The two sub-species of the white rhinoceros were recently re-evaluated and it was suggested to consider them as two distinct species – Northern white rhinoceros (*Ceratotherum cottoni*) and Southern white rhinoceros (*C. simum*) (Groves et al. 2010). Inbreeding is a problem that affects semen quality as well. An inverse correlation was found between the level of inbreeding and the quality of the semen collected from populations of three endangered gazelle species (*Gazella dorcas, Gazella dama* and *Gazella cuvieri*) (Gomendio et al. 2000). Poor semen quality was also found in the clouded leopard (*Neofelis nebulosa*) in North American zoos whose population there originated from a founding group of 7 individuals (Pukazhenthi et al. 2006). In this population, close to

90% of the spermatozoa were with abnormal, damaged acrosomes. A similar low quality sperm in association with poor gene diversity was also reported in the Florida panther (*Puma concolor coryi*) (Barone et al. 1994; Holt and Pickard 1999; Roelke et al. 1993).

SPERM CRYOPRESERVATION

To start with, sperm cryopreservation in endangered species is not an easy undertaking. For most species, there is little or no available information about the various factors associated with the process. Does the seminal plasma need to be removed by centrifugation? Which diluents and cryoprotectants will be most suitable and at what concentrations, osmolarity and pH? Are the cells chilling-sensitive? Which freezing method should be used? And how to thaw the samples? Should the cryoprotectants be removed or diluted after thawing? To determine all these and many other questions, basic research is needed for each species under evaluation. Since the first report on semen cryopreservation more than 60 years ago, which resulted from a chance observation (Polge et al. 1949), the field of sperm cryopreservation made progress primarily through trial and error. Even today, after six decades of intensive studies propelled primarily by research related to human infertility and livestock and laboratory animal production, our knowledge about the exact mechanisms that eventually lead to success, failure or anywhere in-between is still very limited (Saragusty et al. 2009a). Describing the advances made in each endangered species is beyond the scope of this report. We will concentrate here on recent advances related to the cryopreservation and thawing/warming processes and point out some unique aspects of them.

The first step towards cryopreservation would be to determine the composition of the cryopreservation extender. The four basic components of the extender are a buffer, a cryoprotectant, a source of energy and often a source of lipids. Alterations in the buffer composition can be made to change the osmolarity and pH of the solution. Cryoprotectants are divided into cell permeating and non-permeating materials. Among the permeating ones, glycerol is probably the most widely used. While spermatozoa of some species tolerate high glycerol concentrations, those of others can survive only very low concentrations. Spermatozoa of marsupials such as the koala (*Phascolarctos cinereus*) or the kangaroo (*Macropus giganteus*) freezes in concentrations as high as 18% or more (Holt and Pickard 1999; Johnston et al. 1993), whereas spermatozoa of other species such as the mouse or pig can tolerate only very low glycerol concentrations. In some cases, like we have found recently in the Asian elephant, gradual addition of the glycerol with the bulk of it added shortly before freezing, can minimize its damage while utilizing its protective affects (Saragusty et al. 2009e). Other permeating cryoprotectants include dimethyl sulfoxide (Me_2SO; a.k.a. DMSO), ethylene glycol and propylene glycol. Non-permeating cryoprotectants are usually di- or oligosaccharides (lactose, raffinose, trehalose, sucrose) or polymers such as polyvinylpyrrolidone (PVP). The concentration to be adopted for any of these to achieve optimal results is a kind of delicate balance between the toxicity of the cryoprotectant and its protective affect. To avoid the risk of toxicity, attempts were done to cryopreserve mouse sperm with no permeating cryoprotectant at all. Sperm cryopreserved in EGTA-Tris-HCl buffered solution produced similar fertilizing ability to fresh semen when used in ICSI, and developed to viable fetuses when transferred embryos were at the 2-cell stage (Ward et al. 2003). The presence of the cryoprotectant may hinder fertilization in some animals so it should be removed after thawing and before AI. In the Poitou donkey (Baudet Du Poitou, an endangered breed of the domestic donkey *Equus asinus*), pregnancies were achieved only if the 4% glycerol was removed from the thawed samples before the jennies were inseminated (Trimeche et al. 1998; Vidament et al. 2005).

Figure 3. Acacia gazelle (*Gazella gazella acaciae*), an endangered gazelle subspecies, is one of the models in which we have demonstrated post mortem semen collection and cryopreservation in large volumes by the directional freezing technique (Saragusty et al. 2006). Photo by Eyal Bartove ©.

The phospholipids of mammalian spermatozoa possess a distinctive and highly unusual fatty acid composition, the most unique feature of which is a very high proportion of long chain (C20-22) highly polyunsaturated fatty acyl groups (Rooke et al. 2001). In most mammals, docosahexaenoic acid (DHA; 22:6,n-3) is the dominant polyunsaturated fatty acid (PUFA) although, in several species, docosapentaenoic acid (22:5,n-6) is also a major component (Rooke et al. 2001). Lipid composition is a major determinant of the membrane flexibility required for the characteristic flagella movement of spermatozoa and for the fusogenic properties of the membranes, associated with the acrosome reaction and fertilization (Langlais and Roberts 1985; Papahadjopoulos et al. 1973). During the process of cryopreservation, the cells go through two major stages - they are chilled from body temperature down to about 4°C and they are then cryopreserved to cryogenic temperatures. In the first stage the chilling process causes changes in the consistency of the cellular plasma membrane, which is called lipid phase transition. These changes result in the transition of membrane lipids from a liquid state to a gel-like state. The lipid phase transition temperature will be low when: 1) the lipid chain is unsaturated and the higher the degree of unsaturation the lower the temperature would be, and 2) the lipid chain is short. This means that during chilling, the long, saturated fatty acids would form raft-like structures that drift in the still fluid unsaturated and short-chain lipids, a process known as lipid phase separation. These membrane changes are damaging, resulting in chilling injury and increased membrane permeability (Arav et al. 2000a; Arav et al. 1996). The ability of the sperm plasma membrane to resist structural damage during cryopreservation may therefore be related to its fatty acids composition and the strength of the bonds between membrane components (Hammerstedt et al. 1990). Not all animals' sperm are sensitive to chilling. It was shown, for example, that spermatozoa of Asian elephants, which are more chilling sensitive than African elephant sperm, contain lower concentration of PUFA in their membranes and are especially poor in DHA (Swain and Miller 2000). Analysis of membrane lipid composition (Drobnis et al. 1993) or the use of Fourier transform infrared spectroscopy can help in identifying the temperature range at which species-specific spermatozoa are most sensitive, as we have shown, for example, in Asian elephants (Saragusty et al. 2005).

There are several ways to influence the lipid composition of the sperm membrane. Most sperm freezing extenders contain egg yolk or other lipid source. We have shown that when Asian elephant spermatozoa were incubated in extender containing egg yolk or even only with liposomes made out of egg-phosphatidylcholine, the lipid phase transition temperature was down-shifted by more than 12 degrees, making it much less sensitive to chilling injury (Saragusty et al. 2005). Others showed that the addition of cholesterol also increases membrane fluidity and provide similar protection (Moore et al. 2005; Purdy and Graham 2004). Various reports on studies done by us and others show that enrichment of the diet with PUFA can improve motility and concentration, alter the biophysical properties and enhance the reproductive performance of mammalian and avian sperm (Blesbois et al. 1997; Brinsko et al. 2003; Culver 2001; Gacitua et al. 2002; Mitre et al. 2004; Strzezek et al. 2004).

While long chain poly- or monounsaturated fatty acids seem to be beneficial to sperm motility, such fatty acids are also highly prone to oxidation, resulting in reactive oxygen species (ROS), which are known to be damaging to both sperm plasma membrane and DNA (Baumber et al. 2000; Chen et al. 2002). In an attempt to prevent these damaging effects, numerous studies were conducted using various antioxidants (vitamin C, vitamin E, glutathione, □-carotene to name a few), showing beneficial effect of these compounds on sperm characteristics (Ball et al. 2001; Comhaire et al. 2000). These antioxidants are often added to the cryopreservation extender to protect the cells during processing and after thawing.

One of the questions that should be answered is whether it is necessary to remove the seminal plasma before suspending the sperm in the cryoprotective extender. In some mammals such as the goat or the Spanish ibex (*Capra pyrenaica*) (Coloma et al. 2010), the seminal plasma contains phospholipase enzymes that react with components in the egg yolk, leading to severe damage to the cells in the sample. Yet, in other mammals, such as camelids or Kuala, the seminal plasma contain ovulation-inducing factor, rendering it essential during AI (Adams et al. 2005; Johnston et al. 2004; Pan et al. 2001). If the seminal plasma needs to be removed, we found that cushioned centrifugation was beneficial in terms of total sperm yield and centrifugation force damage reduction (Revell et al. 1997; Saragusty et al. 2006; Saragusty et al. 2007). In this process, the semen sample is underlain with an inert, non-ionic, high-density fluid such as 60% aqueous iodixanol solution. At the end of centrifugation, the sperm pellet is at the interface between the iodixanol and the seminal plasma. Iodixanol and the seminal plasma can then be aspirated. With iodixanol, a higher centrifugation force can be used to increase the yield. We have also showed that in addition to the benefits of iodixanol during centrifugation, it also seem to provide protection to the cells during cryopreservation, most probably by increasing the extender's glass transition temperature and by altering the ice crystal morphology, thus making them more hospitable to the cells (Saragusty et al. 2009b; Figure 4). The presence of seminal plasma, however, may confer some protection to the cells after thawing and enhance sperm function as was shown for instance in both boar and stallion (de Andrade et al. 2010; Okazaki et al. 2009).

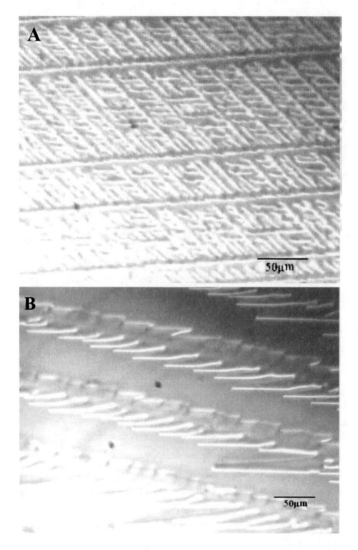

Figure 4. Ice crystal morphology during cryopreservation on a directional freezing cryomicroscope stage when (A) the freezing extender contained 2.5% (v/v) iodixanol solution, and (B) in the absence of iodixanol.

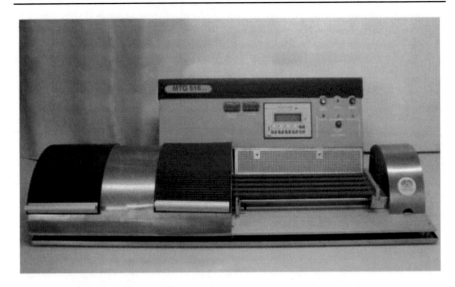

Figure 5. Multi-Thermal Gradient (MTG) device in which large-volume test tubes with samples are advanced at a constant velocity through a pre-defined temperature gradient to achieve optimal heat transfer and extra-cellular ice crystallization.

Most cryopreservation protocols call for slow and gradual chilling of the extended sperm sample down to the pre-freezing temperature of ~4°C. This slow chilling provide the sperm with ample time for equilibration and thus to a reduced chilling injury. Recently a report on human spermatozoa suggested that diluting the semen with cold (4°C) extender, and thus facilitating fast chilling, resulted in an improvement of 15% in post-thaw sperm motility (Clarke et al. 2003). Despite previous reports that fast chilling rates are damaging to sperm cells (Fiser and Fairfull 1986) and the fact that it is known that Asian elephant spermatozoa is sensitive to chilling (Saragusty et al. 2005), we tested this method of cold extender addition on Asian elephant sperm. Our results showed that the use of cold extender was clearly damaging to Asian elephant sperm cells (Saragusty et al. 2009e). Thus, it can only be assumed that such an improvement can be achieved in some mammals but not in others, strengthening the notion that any stage in the process needs to be customized for each species.

One treatment that has been proposed as a mean of enhancing germplasm ability to withstand the stresses they are exposed to during the freeze-thaw cycle is their exposure to high hydrostatic pressure just prior to cryopreservation. The application of high hydrostatic pressure to gametes and embryos at a level of 20-90 MPa (200 to 900 times the atmospheric pressure), seem to benefit their cryosurvival. The level of pressure and its duration depend on the species and the type of gamete or embryonic developmental stage. For example, porcine oocytes optimally withstand pressure of only 20 MPa whereas mouse blastocysts can survive pressure as high as 90 MPa for 30 min and then recover to the same level as the control (Du et al. 2008; Pribenszky et al. 2005). Porcine oocytes do not survive a much lower pressure of 60 MPa (Pribenszky et al. 2008). Such improved survival was demonstrated, for example, in pig and bovine oocytes (Du et al. 2008; Pribenszky et al. 2008), mouse blastocysts (Pribenszky et al. 2005) and boar spermatozoa (Pribenszky et al. 2006). This technique was initially demonstrated by Pribenszky et al. (2005) and Du et al. (2008) who suggested that the pressure put the cells under stressful conditions that lead them to produce and accumulate chaperone proteins such as heat shock proteins (HSPs). These proteins seem to be beneficial to the cells during cryopreservation, which is also a stress-inducing procedure. In one study for example, boar semen was exposed to pressure levels ranging between 10 and 80 MPa for 40, 80 and 120 min before chilling and freezing. It was found that exposure to 20 to 40 MPa for 80 min significantly protected sperm motility during chilling and after the freeze-thaw cycle (Pribenszky et al. 2006). After 5 h of chilling there was no decrease in sperm motility if the sperm was exposed to pressure between 10 and 40 MPa but was significantly reduced in the control and the 80 MPa groups. After thawing, sperm motility was 43.2 ± 5.24% and 42.0 ± 3.24% in the 20 and 40 MPa for 80 min groups and only 23.2 ± 1.83% in the control.

Several cryopreservation methods are in use for preservation of spermatozoa. Probably the most widely used one is the liquid nitrogen vapor freezing method during which the extended semen, usually packaged into 0.25 or 0.5 mL plastic straws, is held at a predetermined distance above liquid nitrogen for several minutes before being plunged into the liquid nitrogen for storage (Roussel et al. 1964; Sherman 1963). The distance between the sample and the liquid nitrogen determines the cooling rate and the final temperature reached before transferring the sample into the liquid nitrogen. The optimal distance can thus vary considerably between species. When freezing rhinoceros semen, for example, we found that a distance of just 2 cm gave superior results at any greater distance (unpublished data). A variation on this theme is a stepwise liquid nitrogen vapor freezing technique. Using this method the sample is first placed at a larger distance (higher subzero temperature) for a specific duration of time, then transferred to a shorter distance (lower temperature) and only then into the liquid nitrogen. This technique was used, for instance, to freeze the Namibian cheetah (*Acinonyx jubatus*) spermatozoa (Crosier et al. 2006). Another method is known as the pellet method in which a small volume (usually around 200 μL) of extended semen is placed directly on carbon dioxide ice ("dry ice") and then stored in liquid nitrogen (Gibson and Graham 1969). Other low-tech techniques include the dry-shipper freezing technique (Roth et al. 1999) or freezing in cold ethanol (Saroff and Mixner 1955). In more recent years, controlled rate freezing machines started penetrating the market (Landa and Almquist 1979). In these machines, suitable for freezing extended semen packaged in straws, the rate of both chilling and freezing can be programmed and precisely controlled. The drawback of the various conventional freezing methods (vapor, pellet, controlled rate) is that ice crystal growth is uncontrolled in terms of both velocity and morphology, and the crystals may therefore disrupt and kill cells in the sample (Watson 2000).

The alternative technique, which was described in a recent review as "the only recent significant advancement in semen preservation technology" (Gillan et al. 2004) is the directional freezing technique. This technique uses the Multi-Thermal Gradient device (MTG®; IMT Ltd, Ness Ziona, Israel) (Figure 5). After the initial seeding stage, the semen sample is advanced at a constant velocity through a linear temperature gradient. Ice crystal propagation can thus be controlled, to optimize crystal morphology, and to achieve continuous seeding and a homogenous cooling rate throughout the entire freezing process. Damage to cells, even when freezing in large volumes of 8 mL, is thereby minimized (Arav 1999; Arav et al. 2002a; Arav et al. 2002b; Saragusty et al. 2009a). A large body of studies, conducted by us and others, using this technology, demonstrated the viability and fertilizing ability of frozen–thawed semen from a variety of species such as bovine (Arav et al. 2002b; Hayakawa et al. 2007; Saragusty et al. 2009c), goat buck (Gacitua and Arav 2005), elephant (Hermes et al. 2003; Saragusty et al. 2009e), bottlenose dolphin (O'Brien and Robeck 2006), Beluga whale (O'Brien and Robeck 2010a; Robeck et al. 2010), rhinoceros (Hermes et al. 2009c; Reid et al. 2009), gazelles (Saragusty et al. 2006), equine (Rubei et al. 2004; Saragusty et al. 2007), European brown hare (Hildebrandt et al. 2009), rabbit (Si et al. 2006), rhesus macaque (Si et al. 2009; Si et al. 2010), and common hippopotamus (Saragusty et al. 2010b). The advantages of large volume freezing are apparent when one has to consider the storage space and costs of a large number of samples under liquid nitrogen over extended period of time. In studies conducted on bovine bull semen, we have demonstrated that semen frozen this way can be thawed, packaged in insemination-dose straws, re-frozen and later thawed and used for AI with acceptable fertility results, similar to those achieved in the conventional single-freezing method (Arav et al. 2002b; Saragusty et al. 2009c). A study on rabbit semen, which also utilized the directional freezing technique, found no difference in fertility rate between fresh semen (87.5%) and semen frozen once (73.9%) and a lower but still acceptable rate for semen that was frozen twice (28.6%) (Si et al. 2006). Several studies on human sperm showed that some cells preserve their functional capacity even after several freeze-thaw cycles and can be used to fertilize oocytes through ICSI (Bandularatne and Bongso 2002; Polcz et al. 1998; Rofeim et al. 2001; Verza et al. 2009). Recently it was also shown that if the sperm are kept in the same cryoprotectant solution between freeze-thaw cycles (rather than washing and re-suspending them in a fresh extender) it significantly enhances their ability to withstand the repeated treatment and helps protect their DNA from fragmentation (Thomson et al. 2010).

Another method, which is similar to freezing but very different from it, is vitrification. Ice crystals, both outside the cells and even more so – inside them, can be very damaging to any frozen tissue or cell. Vitrification, also known as ice-free cryopreservation, is a process in which liquid is transformed into an amorphous, glass-like solid, free of any crystalline structures (Luyet 1937). A major advantage of vitrification over the slow freezing technique is its low-tech, low cost, simple-to-use, suitable-for-the-field character. It does, however, require much experience to load the sample on the carrier. To achieve a vitrified state, either very high concentrations of cryoprotectants or very high cooling rates are needed. Since the high cryoprotectant concentrations needed are beyond what spermatozoa can tolerate, the frozen volume is being considerably reduced to achieve high cooling rate with adequate heat transfer throughout the sample. The technique, though, have several major drawbacks when sperm banking is considered: 1) the small volume that can be vitrified (at best, only a few microliters of semen suspended in vitrification solution) is way too little for most practical purposes. The reason for these small volumes is that spermatozoa are very sensitive to the vitrification solutions so volume is decreased to achieve probability of vitrification and also higher cooling rate and thus enable reduction of cryoprotectant concentrations to a non-toxic level. Furthermore, while oocytes and embryos are large and can be easily picked and transferred from one solution to another during the process of exposing the cells in a step-wise manner to vitrification solutions with increasing osmolarity, this can not be done fast enough with spermatozoa, 2) the risk of contamination through the liquid nitrogen as currently many of the available vitrification carriers are open systems, exposing the sample to the liquid nitrogen in order to increase cooling rate, 3) the high permeable cryoprotectant concentrations (up to 40% v/v compared to 3 to 7% in slow freezing) which are still needed despite the reduction in volume. Such concentrations are both toxic and cause osmotic damages to the cells. To overcome this last obstacle, permeating cryoprotectant-free vitrification was attempted. In a series of three studies conducted on human sperm, permeating cryoprotectant-free (but with some sugars and 10% bovine serum albumin) vitrification was achieved by loading the sperm suspension into a copper loop to form a thin film. The copper loop was then plunged into liquid nitrogen to achieve cooling rates as high as 72,000°C per minute. To match the ultra-fast cooling rate, thawing was also done in a very fast warming rate. Under these conditions, researchers were able to achieve with vitrification similar post-warming motility, normal morphology and fertilizing ability as compared to the standard liquid nitrogen vapor freezing with glycerol (Chang et al. 2008; Isachenko et al. 2003a; Isachenko et al. 2004; Nawroth et al. 2002). In a more recent study, these authors showed that even vitrification of 30 µL of sperm, suspended in human tubal fluid containing 1% human serum albumin and 0.25M sucrose, can be done by dropping it into liquid nitrogen. After such treatment, post-warming forward motility was 57.1% (Isachenko et al. 2008). Going to even larger volumes, in a study on rabbit sperm cryopreserved without permeating cryoprotectants, volumes of 100 µL were plunged into the liquid nitrogen (Li et al. 2010). Sperm cryopreserved this way was able to direct full-term development of embryos, which were produced by ICSI.

In human medicine, couples with male factor infertility represents 30-40% of the infertile population and in about 10% of these couples, azoospermia is observed. In many animal species only very small number of spermatozoa can be retrieved either because of male factor infertility or because only small number of cells can be collected. For instance, in the naked mole rat only 5 to 10 µL of sperm can be collected by electroejaculation, and the number of cells in the collection is often very small and many of them seem to be morphologically damaged (unpublished data). In many cases, both in human medicine and wildlife conservation, obtaining ejaculated sperm is not successful or not practical and invasive techniques such as aspiration of epididymal sperm or testicular sperm extraction (TESE) are employed, frequently resulting in a small number of cells. Since the development of the intracytoplasmic sperm injection technique, single sperm can be used to fertilize an oocyte *in vitro*. It would thus be desirable to have at your disposal techniques to also cryopreserve single or small number of spermatozoa, which can later be used in ICSI. Cryopreservation of single or small number of spermatozoa can be challenging for several reasons. From a technical aspect, it might be difficult to locate the cell in the vast expanses of the freezing extender after thawing so naturally one should aim at very small volumes. When going down in volume, the probability of vitrification increases. However, spermatozoa of many species cannot withstand the toxicity of the high concentration of permeating cryoprotectants needed. In some species cryoprotectant-free vitrification has been reported (Isachenko et al. 2003a; Isachenko et al. 2004) but this probably cannot work for many others. Starting in the late 1990's, reports on several single sperm cryopreservation techniques showed up in the scientific literature. The first report suggested using empty (oocyte-free) zona pellucida into which the single sperm was inserted and then cryopreserved (Cohen and Garrisi 1997; Cohen et al. 1997). Human, mouse or hamster zonae pellucidae were all used for this purpose (Borini et al. 2000; Cohen et al. 1997; Hsieh et al. 2000; Walmsley et al. 1998). This technique proved successful and pregnancies and births resulting from sperm cryopreserved this way were reported (Fusi et al. 2001; Walmsley et al. 1998). Improvements to the technique were later reported. In one, the zona pellucida was perforated with the aid of laser (Montag et al. 1999). Through this perforation, the zona pellucida is emptied and the sperm is inserted. In another, the cryoprotectant solution was modified and the new solution, named 'Osmangazi-Turk solution', enhanced sperm survival (Hassa et al. 2006). Following the success with the zona pellucida technique, a number of other cryopreservation techniques were suggested. These included pellets on dry ice where pregnancies were achieved (Gil-Salom et al. 2000), freezing in the ICSI pipette (Gvakharia and Adamson 2001), freezing by placing a single or multiple cryoprotectant microdrops with the sperm under paraffin oil in a culture dish or ICSI dish and then freezing it (Bouamama et al. 2003; Hu et al. 2010; Sereni et al. 2008), using cryoloop for slow freezing (Desai et al. 2004a; Desai et al. 2004b; Schuster et al. 2003) and for vitrification (Isachenko et al. 2004), freezing inside *Volvox globator* spheres (Just et al. 2004), freezing in alginic acid capsules (Herrler et al. 2006), or inside agarose gel microspheres (Isaev et al. 2007), freezing in straws (Koscinski et al. 2007), and in CellSeal[TM] closed system (Woods et al. 2010). Using these various techniques, researchers reported a wide range of outcomes in many of the *in vitro* parameters measured including post-thaw motility, recovery rate, viability, fertilization rate, cleavage rate and blastocyst rate. In the largest (human) study to date that has also evaluated embryo transfer, 156 ICSI cycles with fresh sperm were compared to 234 cycles using frozen-thawed testicular sperm (Gil-

Salom et al. 2000). Overall sperm survival rate, regardless of the etiology of male infertility, was 94%. While the mean embryo cleavage rate was higher in the fresh sperm group (90.6% vs. 84.6%, respectively), fresh and frozen-thawed sperm were similar in mean fertilization rate (62.0% vs. 63.2%, respectively), clinical pregnancy rate per cycle (28.2% vs. 27.8%, respectively), implantation rate (12.2% vs. 13.1%, respectively) and on going pregnancy per cycle (22.4% vs. 21.8%, respectively). In this study, a total of 55 babies were born following cryopreservation of single or small number of sperm and several pregnancies were ongoing. Time will tell which of these technologies, or others that are currently under development, will emerge as the leading technique that will gain a foothold in sperm banks and IVF centers around the world.

What unifies all these cryopreservation methods is that the samples, once cryopreserved, are stored under liquid nitrogen till they are thawed. Liquid nitrogen, however, was shown in several studies to be a medium through which contamination can be transmitted from one sample to another (Bielanski et al. 2003; Bielanski and Vajta 2009; Russell et al. 1997). Three options are possible, then. The first is to test each sample for a wide variety of possible contaminants and use it only if proved negative (Tedder et al. 1995). A second option is to seal each sample so as to prevent possible transmission (Chen et al. 2006b). A third option is to store the samples in the liquid nitrogen vapor phase rather than in the liquid nitrogen itself. In a recent study it was demonstrated that there was no difference between samples stored in liquid nitrogen and those in liquid nitrogen vapor at temperature up to about -160°C for periods of up to 3 months (Lim et al. 2010).

LIQUID PHASE PRESERVATION

In many cases, sperm can be collected in the field, away from any fully equipped andrology and cryobiology laboratory, and transferring the samples to a facility for processing may take time. For such cases, or when the sample is destined to be used but not immediately, mid-term non-freezing preservation techniques may help. In some species, such as the pig, chilled storage (usually about 15 to 17°C) is the most widespread method for sperm preservation as thus far sperm cryopreservation has provided only mediocre post-thaw results. Preliminary experiments conducted in our laboratory showed that using large volume directional freezing technique, boar semen can be frozen with as high as 75% post-thaw motility, suggesting that this technique might be successful where the conventional methods have thus far failed (unpublished data). Boar semen can usually be stored at 15 to 17°C for several days. When planning on extended chilled storage, several sperm energy-metabolism aspects should be taken into consideration. Both glycolysis and the Krebs cycle play an important role in sperm energy metabolism. Besides monosaccharides, spermatozoa use other substrates such as citrate and lactate to produce energy as well. A fine balance between these metabolites is required to produce an optimal preservation solution (Rodriguez-Gil 2006). For mid-term non-freezing preservation, two methods were recently proposed. After evaluating various solutions, osmolarities and storage temperatures, Van Thuan and colleagues (2005) found that the optimal conditions for mouse spermatozoa preservation were in 800 mOsm/kg potassium simplex optimized medium, containing amino acids and 4 mg/mL bovine serum albumin, and a holding temperature of 4°C. More than 40% of oocytes injected with sperm heads stored under these conditions for 2 months developed to the morula/blastocyst stage *in vitro* and 39% of the embryos developed to term after being transfer to recipient mice. Using the same medium, these researchers also showed that mouse spermatozoa can retain competence for ICSI if they are held for one week at room temperature and then for up to 3 months at -20°C. In another study, conducted on human and mouse spermatozoa, the cells were stored at 4°C in electrolyte-free solution containing glucose and bovine serum albumin. Under these conditions, cells preserved their motility for 3 weeks (mice) or 4 weeks (human) and their viability for up to 6 weeks. Both mouse and human spermatozoa stored under these conditions were capable of fertilizing oocytes when used in ICSI (Kanno et al. 1998; Riel et al. 2007). Another alternative is to simply leave the spermatozoa inside the epididymis and keep these at 4°C in isotonic saline solution. Spermatozoa were shown to preserve motility and viability for 8 days this way (Yu and Leibo 2002). However, motility preservation for several days can also be done with ejaculated sperm in egg yolk based extenders. For instance, we have recently showed that pygmy hippopotamus (*Choeropsis liberiensis*) spermatozoa preserved some motility for 3 weeks when suspended in the Berliner Cryomedium (BC) basic solution (a TEST-egg yolk based extender) and stored at 4°C (Saragusty et al. 2010a). Naturally one should keep in mind that during extended storage the solution in which the cells are suspended become increasingly toxic to the surviving spermatozoa because of the release of reactive oxygen species and disintegration of dead cells.

PRESERVATION OF EARLY DEVELOPMENTAL STAGES

Spermatozoa, however, can only be potentially retrieved from adult, relatively healthy, individuals but not from sick, azoospermic or prepubertal ones, and often these carry valuable genetic material that, if not preserved, will be lost for the population. Thanks to ICSI, even early developmental stages such as elongating or elongated spermatids can be utilized for fertilization. Such cells as testicular spermatozoa and earlier developmental stages can be extracted using testicular sperm extraction (TESE) techniques and then used through ICSI to fertilize oocytes (Devroey et al. 1995; Hewitson et al. 2002; Kimura and Yanagimachi 1995; Schoysman et al. 1993). These cells, however, do not direct oocyte activation when injected and this should be done, either before or after sperm injection, to facilitate embryonic development (Kimura and Yanagimachi 1995). These early-stage cells can be used as fresh cells but they can also be cryopreserved and used at a later stage when needed (Hirabayashi et al. 2008).

Cells of even an earlier stage than the spermatocytes and spermatids are the spermatogonium or spermatogonial stem cells which can be collected from any male, including infants and juveniles. Infant mortality rate is known to be relatively high in many captive and wild populations (e.g. Howell-Stephens et al. 2009; Saragusty et al. 2009d) so methods to preserve germ cells from valuable individuals, be it stillborn ones or those who died early in life, is certainly called for. Spermatogonial stem cells transplantation was first reported in mice (Brinster and Zimmermann 1994) when it was demonstrated that such transplantation can lead to spermatogenesis. The transplantation technique was later extended to other, larger, species such as pigs (Honaramooz et al. 2002), bovine (Izadyar et al. 2003) goats (Honaramooz et al. 2003a; Honaramooz et al. 2003b), and cynomolgus monkeys (Schlatt et al. 2002a). Xenogeneic transplantation, usually from other mammals to nude, immune-deficient mice, has also been reported. However, the further phylogenitically apart the donor and recipient species are, the more difficult it becomes. Using this technique, isolated donor testicular cells are infused into the seminiferous tubules of the recipient whose testis has been depleted of all its germ cell line (by irradiation or chemotherapy). Spermatogonial stem cells establish themselves in the testis and through spermatogenesis, produce spermatozoa carrying the donor's genetic material. In 2006 the proof that such xenotransplanted cells can actually produce normal, functioning spermatozoa was reported (Shinohara et al. 2006). In their study, spermatogonial stem cells collected from immature rats were transplanted into chemically sterilized mice and the spermatozoa or spermatids collected from the recipient mice produced normal, fertile rat offspring both when freshly used and following cryopreservation. The demonstration that such stem cells can be collected, transplanted and produce viable spermatozoa indicate that this is yet another useful way for the conservation of genetic material, for example from immature deceased animals. Donor stem cells can be grown in culture to generate more cells for transplantation (Nagano et al. 1998) and they can be cryopreserved for future use (Avarbock et al. 1996). Under very complex *in vitro* culture conditions, and with very low efficiency, morphologically normal and even motile spermatozoa were generated from spermatogonial stem cells (Feng et al. 2002; Hong et al. 2004; Stukenborg et al. 2009). The demonstration that these cells can be reproductively viable came very recently when reproductively healthy mice were produced following ICSI with spermatozoa retrieved from neonatal testicular tissue that was cultured for two months for spermatogenesis to be completed *in vitro* (Sato et al. 2011).

TESTICULAR TISSUE PRESERVATION

Tissue cryopreservation is more complex than cellular preservation because tissue is composed of more than one cell type and thus of different water and cryoprotectant permeability coefficients and different sensitivities to chilling and osmotic challenges. Tissue is also larger in volume and thus cryoprotectant penetration is difficult and heat transfer is not uniform, putting the center of the sample at greater risk of intracellular ice formation and/or recrystallization and death. This is true for testicular tissue, ovarian tissues or any other tissue type. Testicular tissue preservation can basically be done in two forms. Either the tissue is cryopreserved for future use or it is transplanted. When preserved in the frozen form, both spermatozoa and spermatogonial stem cells can be harvested from the tissue after thawing. Recently it was demonstrated that spermatozoa or spermatids retrieved from reproductive tissues (whole testes or epididymides) which was stored frozen for up to one year at -80°C or from whole mice kept at -20°C for up to 15 years can produce normal offspring following ICSI (Ogonuki et al. 2006). This success followed a previous, failed, attempt to cryopreserve the entire testis (Yin et al. 2003). The other preservation alternative is testicular tissue cryopreservation. The technique of testicular tissue cryopreservation is widely used today in both adult and pediatric human medicine as a mean to preserve spermatozoa or spermatogonial stem cells from patients undergoing cancer treatments. To cryopreserve the tissue, it is cut into tiny pieces, usually in the range of 1 to 2 mm^3 to ensure cryoprotectant penetration, efficient heat transfer and eventual successful grafting. Other alternatives that have been proposed are to mince the tissue and then suspend it in freezing extender to achieve better cryoprotection (Crabbe et al. 1999) or to cut the testicular tissue into thin stripes (e.g. ~9×5×1 mm as was done in sheep) to increase the total number of seminiferous tubules in each graft (Rodriguez-Sosa et al. 2010). Although such tissue samples can be obtained from every individual, be it infant, juvenile or adult, almost all successful studies to date used immature tissue (Ehmcke and Schlatt 2008). Like in semen freezing, there are differences between species in the reaction of their testicular tissue to cryoprotectants, chilling and freezing (Schlatt et al. 2002b). The preserved testicular tissue can be handled in several ways. From these tissues, spermatozoa, spermatocytes and round and elongated spermatids can all be retrieved and used in ICSI (Gianaroli et al. 1999; Hovatta et al. 1996). Frozen-thawed testicular tissue can be transplanted back to the donating individual (autografting), to another individual of the same species (allografting) or to individual of a different species, usually to nude or immunodeficient mice (xenografting). After transplantation, many grafts are lost due to tissue rejection or ischemia. If it survived the critical first few days, blood supply will reach the graft, it will be supported by the recipient metabolic and hormonal system and, after some time, will start producing spermatozoa, which can be harvested by surgical excision of all or part of the graft (Schlatt et al. 2002b). Although dependent on the recipient system for support, the spermatogenesis cycle length is assumed to be inherent to the spermatogonial stem cells, which are expected to preserve the donating species spermatogenesis length (Zeng et al. 2006). However, some studies showed that in a few species, the process is accelerated when their testicular tissue was xenografted into mice (Honaramooz et al. 2004) while in others it was not (Snedaker et al. 2004). Acceleration, when identified, bears special interest for species preservation as it can shorten generation time and thus speed up population growth. Sperm produced this way does not go through epididymal maturation process so the only way it can be utilized is by ICSI (Shinohara et al. 2002), a technique (along with its associated procedures – *in vitro* embryo culture and embryo transfer) not yet developed for most species. One should also keep in mind that it is very costly to keep immunodeficient mice and handle them under germ-free conditions and, of course, repeated transplantations from one mouse to another are required to maintain viable tissue for many years. Still, testicular tissue cryopreservation was done in several species and pregnancies were achieved in mice (Schlatt et al. 2002b; Shinohara et al. 2002), rabbit (Shinohara et al. 2002), human (Hovatta et al. 1996), Djungarian hamsters (Schlatt et al. 2002b) and marmoset monkeys (Schlatt et al. 2002b) to name a few.

SEX-SORTING OF SPERM

An important procedure that holds great potential for wildlife conservation is the pre-selection of offspring gender through the use of spermatozoa that were sorted according to the sex chromosome they carry. This technique is based on the small, but significant, difference in DNA content between the X- and Y-chromosome bearing spermatozoa that exist in many (but not all) species (Johnson 1988; Johnson et al. 1987). A cell sorting machine, specifically modified for this purpose, can separate the sample into three populations - X-chromosome bearing spermatozoa, Y-chromosome bearing spermatozoa and unsortable cells (dead or maloriented in relation to the laser beam). The sorting purity can exceed 90%, although this comes at a price of reduced sorting rate and a higher percentage of loss. Tens of thousands of viable, normal offspring were produced thus far, primarily in the livestock industry. Development of the sex sorting technique as part of the ART tool box of wildlife management will improve our ability to better control the sex ratio within social groups, simulating that in the wild and, in endangered species, where the production of a large number of offspring is crucial, produce more females. As a first step, the difference in DNA content between the X- and Y-chromosome bearing spermatozoa should be determined. This was done already for a large number of species (for review see Garner 2006) and attempts at using this technique to sort spermatozoa of wildlife species were done by us and others in non-human primates, alpaca, rhino and elephant (Behr et al. 2009a; Hermes et al. 2009a; Morton et al. 2008; O'Brien et al. 2004; O'Brien et al. 2005a; O'Brien et al. 2005b), with offspring born in elk and dolphins (O'Brien and Robeck 2006; Schenk and DeGrofft 2003). Some difficulties are still associated with this technique, making it hard to utilize its full potential. One problem is the location of such sorting machines in relation to both the sperm donor and the female to be inseminated with the sorted sample. There are only a handful of such sorting machines around the world so the chances they will be in proximity to where they are needed are small. The double freezing technique mentioned earlier may come handy in this respect (Saragusty et al. 2009c). Semen can be cryopreserved, shipped to the sorting machine, thawed, sorted, refrozen in insemination doses and shipped to where it will be used for insemination. This idea was recently demonstrated in bull and ram (de Graaf et al. 2007; Underwood et al. 2007). Another problem is the sorting rate. Under optimal conditions, only about 20 million cells of each sex can be sorted by one machine every hour. Usually the rate is much lower so the number of sorted cells that can be collected within reasonable time is small. While this is less of a problem in animals where small insemination doses are used and when intrauterine insemination is possible, it certainly is a problem in animals, such as the elephant, where large sperm numbers in large volumes are needed for each insemination and the sperm cannot usually be deposited further than the cervix. The cells, after going through all the stresses involved in sorting and post-sorting handling, also suffer from reduced viability over time. And the surviving cells seem to be of reduced fertilizing and developmental quality. For instance, it was reported recently that pregnancy rate following the use of sex-sorted sperm that were frozen after sexing was considerably lower than the control group and almost all pregnancies that were achieve were lost (Underwood et al. 2010). However, a few AI centers in the UK and Japan cryopreserved successfully sorted sperm using the directional freezing technique. This field is still young and progress is fast although most of the studies are restricted to just a handful of research groups that have access to a sorting machine. Future will probably bring the needed technologies that will make things work better for the endangered species that need it so much.

Figure 6. Trans-illumination of this sheep ovary shows many antral follicles. The ovary was isolated from a sheep six years after transplantation of a frozen-thawed ovary (Arav 2001; Arav et al. 2010).

SPERM PRESERVATION - CONCLUSIONS

While the ultimate goal of any of the options described above is to obtain viable, motile, intact spermatozoa after cryopreservation so that they can be used for AI or IVF, since the development of ICSI (Goto et al. 1991), sperm motility became obsolete and in many cases, sonication is used to remove the tails altogether prior to their injection into oocyte. In some species, such as the clouded leopard (*Neofelis nebulosa*), sperm quality is very poor, it hardly produces any embryos during IVF and it does not survive well the cryopreservation process (Pukazhenthi et al. 2006). Such sperm, however, can still be preserved and used in ICSI as part of the conservation efforts of such species. The advent of the ICSI technique also opened the way to a thoroughly different form of sperm preservation – freeze-drying. This technique will be discussed later in this book. The eventual use of preserved spermatozoa, be it for AI, IVF or ICSI, requires both the ability to handle gametes *in vitro* and, in the case of IVF or ICSI, also the ability to obtain and handle oocytes, fertilize them and eventually transfer the resulting embryos, following *in vitro* culture, to recipient females. Naturally, in all these cases, a thorough understanding of the female reproductive biology is a pre-requisite.

Chapter 2

OOCYTES ASPIRATION AND PRESERVATION

Collecting cells from dead animals is a valuable technique for cryopreservation of endangered species. We have shown that it is possible to collect sperm from the cauda epididymis up to 24h after death and to maintain cellular viability after freezing and thawing (Saragusty et al. 2006). For female reproductive preservation, the parallel procedure should be the collection of oocytes at the germinal vesicle (GV) stage, followed by *in vitro* maturation and oocytes vitrification. It was shown by us that oocytes aspiration, done by the trans-illumination technique (Figure 6), increase the number of oocytes collected from isolated ovaries by 50% (Arav 2001). In cows, an average of 7.3 oocytes were collected from each ovary (Arav 2001) and from pigs and sheep the number of GV oocytes that can be collected is even higher. *In vitro* maturation, however, is a complex procedure not yet established for most species and even for the handful of species for which it has been attempted, success is fairly limited (Krisher 2004). Furthermore, collection of immature oocytes disrupts the natural maturation process and thus compromises the quality of the oocytes even if they were later matured *in vitro*. During oocyte maturation and follicular growth, the oocyte accumulates large quantities of mRNA and proteins needed for the continuation of meiosis, fertilization and embryonic development. In the absence of the entire supporting system during *in vitro* culture, the production of some of these needed components is hampered. In seasonal animals, oocytes collected out of the season often show resistance to IVM and IVF (Berg and Asher 2003; Comizzoli et al. 2003; Spindler et al. 2000). In red deer, for example, while about 15% of cleaved oocytes collected during the season (April-July) developed *in vitro* to blastocysts, none have developed if collected after July (Berg and Asher

2003). Comizzoli and colleagues (2003) showed that adding anti-oxidants and FSH into the culture media can help to at least partially overcome this problem in the domestic cat model they were working with. Naturally, *in vitro* fertilization and embryo culture should also be developed so that embryos can be generated for transfer. To enhance the number of oocytes collected at any ovum pick-up procedure, hormonal stimulation can be used. This, however, will result in both mature and immature oocytes and the quality of both may be compromised (Blondin et al. 1996; Moor et al. 1998; Takagi et al. 2001). Although to date no morphological or other method is able to accurately predict which oocytes have optimal developmental potential (Coticchio et al. 2004), it is clear that oocyte quality is a major determining factor in the success of IVF (Combelles and Racowsky 2005; Coticchio et al. 2004; Krisher 2004), early embryonic survival, the establishment and maintenance of pregnancy, fetal development, and even adult disease (Krisher 2004). Once all these hurdles have been overcome and while keeping in mind the importance of oocyte quality, the next major hurdle to surmount is oocyte cryopreservation.

Oocytes are very different from sperm or embryos with respect to cryopreservation. The size of the oocyte is in the range of 3 to 4 orders of magnitude larger than that of spermatozoa, thus substantially decreasing the surface-to-volume ratio, making them very sensitive to chilling and highly susceptible to intracellular ice formation (Chen and Yang 2009; Zeron et al. 1999). However, despite the fact that the surface-to-volume ratio of a 2 PN human oocyte is the same as that of mature oocyte, it is the most freezable stage for the human female gamete (Ghetler et al. 2005). Oocytes are surrounded by zona pellucida, which acts as an additional barrier to the movement of water and cryoprotectants into and out of the cryopreserved oocyte. Oocytes at the MII stage also have a formed fuse that is chilling-sensitive (Chen and Yang 2009) and their plasma membrane has low permeability coefficient, thus making the movement of cryoprotectants and water slower (Ruffing et al. 1993). They also have high cytoplasmic lipid content which increases chilling sensitivity (Ruffing et al. 1993). Oocytes have less submembranous actin microtubules (Gook et al. 1993) making their membrane less robust. The meiotic spindle, which has formed by the MII stage, is very sensitive to chilling and may be compromised as well (Ciotti et al. 2009) and oocytes are more susceptible to the damaging effects of reactive oxygen species (Gupta et al. 2010). Oocytes also present high sensitivity to low temperature upon cooling and before freezing (chilling injury). Chilling injury depend on the biochemical composition and the thermo-behavior of the membranes' lipids (Arav et al. 2000a; Arav et al. 1996; Zeron et al. 2002a; Zeron et al. 2002b). Many of these parameters change after fertilization, making embryos less chilling sensitive and easier to freeze (Fabbri et al. 2000; Gook et al. 1993). Recently we have shown that human oocytes are highly sensitive to cryopreservation while fertilized oocytes are very resistant. The reason is the temperature at which lipid phase transition (LPT) occurs - between 20 and 16°C for mature oocytes and 10°C lower for fertilized oocytes (Ghetler et al. 2005). Chilling injury, which is a kinetic process, is reduced by temperature and exposure time (Zeron et al. 1999). Therefore, LPT that occur at high temperature and for longer time will be more damaging than LPT that occur at lower temperature and for a very short time. We, and others, have shown that rapid cooling rate will overcome (outrace) the problems associated with chilling injury (Figure 7) in a variety of cells models (Arav 1992; Arav et al. 2000b; Mazur et al. 1992).

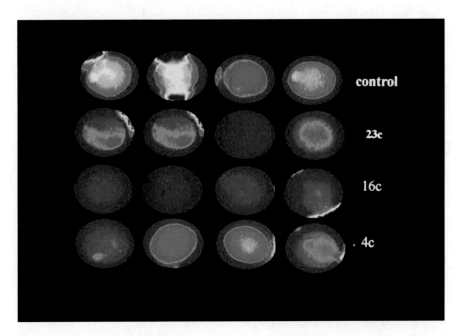

Figure 7. The figure shows bovine oocytes stained with vital dye (cFDA) following rapid cooling (>1000°C/min) and short exposure to different temperatures. Green indicates membrane integrity and blue is for membrane's damage. It is possible to note that 16°C is the most damaging temperature while short exposure to lower temperature is less damaging.

Despite many advances in the field of cryopreservation, oocyte (ovulated, mature or immature) cryopreservation is still not a matured procedure. Even in human medicine, fewer than 200 births resulting from cryopreserved oocytes were reported as of 2007 (Edgar and Gook), a number that went up to around 500 by 2009 (Nagy et al.). Yet, despite all these difficulties, some success in oocyte cryopreservation has been reported. The two basic cryopreservation techniques used are the controlled-rate freezing and vitrification. In controlled rate freezing, oocytes are exposed to permeating cryoprotectants in the range of 1.0-1.5 M and frozen, following equilibration and seeding, at a rate of 0.3-0.5°C per minute down to -30°C or lower, at which point they are plunged into liquid nitrogen for storage. Vitrification exposes the oocytes to substantially higher concentration of cryoprotectants, in the range of 5.0-7.0 M, and cryopreservation is done at very high cooling rates of 2,500°C/min or more (and as high as 250,000°C/min), depending on the method used (Lee et al. 2010; Risco et al. 2007). The first human pregnancy from cryopreserved (by slow freezing), *in vitro* fertilized oocyte was reported in 1986 (Chen) following success in other (laboratory) species, which came a few years earlier, such as the mouse (Whittingham 1977) or the rat (Kasai et al. 1979). Still, despite several decades of research, success is very limited. A meta-analysis on slow freezing of oocytes showed that clinical pregnancy rate per thawed oocytes was only 2.4% (95/4000) and only 1.9% (76/4000) live birth (Oktay et al. 2006).

It was thus proposed that vitrification, following very rapid cooling, would allow cooling to low temperatures without chilling injury. In the year 2008 we celebrated 70 years since the first publication of Luyet on the vitrification of frog sperm (Luyet and Hoddap 1938). Vitrification is a thermodynamic process by which liquid solution will go through solidification without the formation of ice crystals (Luyet 1937). The probability of vitrification depends on three factors: cooling rate, viscosity and volume, thus following the general formula for the probability of vitrification:

$$\mathrm{Pr}\,obability\ of\ Vitrification = \frac{Cooling\,rate \times Vis\cos ity}{Volume}$$

Increasing the cooling rate or the viscosity of the solution or decreasing the volume will increase the probability for vitrification (Arav 1992; Yavin and Arav 2007). Cells at physiological temperatures are in a liquid state and can survive suspended animation (i.e. cryopreservation) ONLY when they are transferred into a solid state (vitrification). Cryopreservation of cells can be achieved using slow freezing, rapid freezing, freeze-drying and vitrification. In any of these methods the cell will survive if it reaches glass transition temperature (Tg) and remains below this Tg for storage, before rapid warming or rehydration (Figure 8). During cooling, homogeneous ice nucleation first occurs between the homogeneous nucleation temperature (Th) and the melting temperature (Tm; e.g. −42°C for water). When the solution viscosity increases (higher cryoprotectant agent concentration) the Th and Tm decrease and Tg increase. By increasing cryoprotectant agent (CPA) concentration, Tg and Tm get closer and eventually meet. As a result, a vitrified solid can be generated even at a slow cooling rate in the case of ~60% (wt/wt) of CPAs (Arav et al. 2002a). Increase of the hydrostatic pressure (over 1000 atm), on the other hand, will result in a decrease in Th and an increase in Tg (Kanno et al. 1975). Using this phenomenon, concentration of cryoprotectants can be reduced. Freezing in closed containers is bound to increase the pressure therein as we have shown recently (Saragusty et al. 2009a).

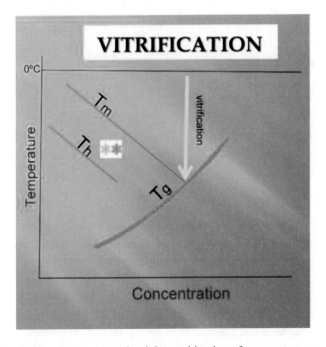

Figure 8. For vitrification to occur, the right combination of cryoprotectant concentration and temperature should be achieved to bring the sample to its glass transition temperature (Tg). Before reaching the Tg, homogenous ice nucleation occurs between the homogenous nucleation temperature (Th) and the melting temperature (Tm).

Table 1. The effect of cooling rate on survival; Comparison between liquid nitrogen and liquid nitrogen slush

Bovine MII	48%	28%	P<0.05	(Arav and Zeron 1997)
Ovine GV	25%	5%	P<0.05	(Isachenko et al. 2001)
Porcine blastocysts	83%	62%	P<0.05	(Beebe et al. 2005)
Bovine MII	48%	39%	P<0.05	(Santos et al. 2006)
Mouse 4-cell with biopsied blastomere	87%	50%	P<0.05	(Lee et al. 2007)
Mouse oocytes	90%	45%	P<0.001	(Lee et al. 2010)
Human oocytes	92%	56%	P<0.05	(Criado et al. 2010)
Rabbit embryos	92%	83%	NS	(Papis et al. 2009)
Porcine blastocysts	89%	93%	NS	(Cuello et al. 2004)
Mouse MII	>80%	>80%	NS	(Seki and Mazur 2009)
Rabbit oocytes	82%	83%	NS	(Cai et al. 2005)

LN = liquid nitrogen, GV = germinal vesicle, Sig. = statistical significance, NS = not significant.

In 1992 we developed the minimum drop size (MDS) technique; reduction of sample volume, which allow reducing the likelihood for crystallization and hence increasing the probability of vitrification (Arav 1992; Arav and Zeron 1997). The use of cryoprotectants (CPAs) with high glass formation properties and high permeability also contribute to the progress in oocytes vitrification (Kasai 2002; Vajta et al. 1998; Yavin and Arav 2007). Vitrification gained a foothold only after 2005, prior to which only 10 human pregnancies resulting from vitrified oocytes were reported (Oktay et al. 2006). Although high oocyte survival rate is achieved with both cryopreservation methods, fertilization rate and embryo transfer rate are still considerably lower than when fresh oocytes are used (Magli et al. 2010). Human oocyte survival rate, using vitrification, is reported to be higher than when using slow freezing [95% (899/948) vs. 75% (1275/1683), respectively] but pregnancy rate per thawed oocyte is still low - in the range of 1.9 to 8.6% for slow freezing and 3.9 to 18.8% for vitrification (Chen and Yang 2009). Even among women with repetitive reproductive success, the rate of live birth per oocyte retrieved was reported to be 7.3% (180/2470) among best-prognosis donors and even lower (5.0%; 52/1044) among standard donors (Martin et al. 2010). Cryopreservation can cause cytoskeleton disorganization, chromosome and DNA abnormalities, spindle disintegration, plasma membrane disruption and premature cortical granule exocytosis with its associated hardening of the zona pellucida (Luvoni 2000). The meiotic spindle, however, tends to recover to some extent after thawing and *in vitro* culture, recovery that is faster following vitrification than following slow freezing (Ciotti et al. 2009). To overcome some of the associated difficulties, ways to increase cooling rate even further were devised. It was suggested that high cooling rate (and warming rate) is expected to decrease crystallization, which is a kinetic process, and takes time to grow. We developed a device, the VitMaster® (Figure 9) which cools the liquid nitrogen (LN) to close to its freezing temperature (between -205 and -210°C). Liquid nitrogen slush, defined as nitrogen in a liquid state close to its freezing point (-210°C), increase the cooling rate by 2 to 5 fold over LN (Yavin and Arav 2007) leading to the decrease in both the probability of crystallization and chilling injury during vitrification. Specific papers that evaluated LN in comparison to LN slush include our reports on bovine oocytes (Arav and Zeron 1997; Arav et al. 2000b) and more recently by Santos et al. (2006), on sheep oocytes (Isachenko et al. 2001), porcine blastocysts (Beebe et al. 2005) and recent publications on manipulated mouse embryos (Lee et al. 2007) and oocytes (Lee et al. 2010). These studies showed 9 to 45% survival improvement when using LN slush (Table 1). Liquid nitrogen slush was also successfully applied to the vitrification of human oocytes (Criado et al. 2010; Yoon et al. 2007).

Immature oocytes seem to be less prone to damages caused by the chilling, freezing and thawing procedures and they, too can be cryopreserved by the controlled rate freezing technique (Luvoni et al. 1997) or by vitrification (Arav et al. 1993; Czarny et al. 2009). Preantral oocytes can be preserved inside the follicle and about 10% seem to be physiologically active after thawing and one week of culture. Of over 16,000 small preantral follicles recovered from the ovaries of 25 cats, 66.3% were intact after thawing (Jewgenow et al. 1997). Before freezing 33.9% of the follicles contained viable oocytes while after thawing there were 19.3% if frozen in Me_2SO and

18.5% if frozen in 1,2-propanediol. However, culture conditions that will allow these oocytes to grow and reach full maturation are still largely unknown despite attempts in several species (Jewgenow et al. 1998; Nayudu et al. 2003). For example, in the marmoset monkey (*Callithrix jacchus*), oocytes collected from secondary pre-antral follicles of either mature or pre-pubertal females were able to develop *in vitro* to the polar body stage but could not complete the maturation process (Nayudu et al. 2003). The exception is the mouse, in which the entire maturation process was successfully performed *in vitro*. Embryos were produced following IVF of frozen-thawed *in vitro* matured primary follicles and live young were born after ET (Carroll et al. 1990). Some, very limited, success was also reported in cats where following vitrification in 40% ethylene glycol, 3.7% of the *in vitro* matured oocytes were able to develop to the blastocyst stage following IVF (Murakami et al. 2004). The problems associated with maturation of early-stage oocytes *in vitro* are the need to develop the complex endocrine system that support the development at different stages, other culture conditions that will ensure survival (oxygen pressure for example) and, in many species, the duration of time required to keep the follicles in culture – 6 months or more. An alternative to isolated oocyte cryopreservation is freezing individual primordial follicles and later transplanting them to the ovarian bursa, where they can mature and eventually produce young offspring following natural mating as was shown in mice (Carroll and Gosden 1993).

Figure 9. The Vitmaster® is a device which cools liquid nitrogen to a temperature close to its freezing point (-210°C) by applying negative pressure.

THE PROTOCOL WE USE FOR OOCYTES VITRIFICATION

Mature oocytes at the MII stage or embryos can be vitrified with the same protocol. Oocytes or embryos are exposed to 10% vitrification solution (VS) for 5-9 minutes, transferred into 50% VS and immediately thereafter into a final 100% VS [containing 38% v/v ethylene glycol (EG), 0.5M trehalose and 6% BSA in PBS). They are then loaded into a tubing carrier or placed onto a surface carrier and vitrified at a rapid cooling rate (CR) of 15,500°C/minute, using the VitMaster® apparatus. The oocytes or embryos are then stored under LN until used. Warming of the oocytes or embryos is performed by plunging the carrier into the warming solution. They are then immersed in 0.6M trehalose solution for 2.5 minutes at 37°C and transferred though a series of solutions containing decreasing concentrations of trehalose: 0.5M, 0.4M, 0.3M, 0.2M and 0.1M for 2 minutes each.

Chapter 3

EMBRYO CRYOPRESERVATION

For probably most of the species on Earth, with the current knowledge in cryopreservation, only the male gametes can be preserved whereas embryos at any stage of development or oocytes cannot. The most prominent difference between spermatozoa and oocytes or embryos is probably the volume. The latter two have volume orders of magnitude larger than the former. As such, the issue of intracellular ice formation is becoming a major concern, even at relatively slow cooling rates. To avoid this from happening small volume cryopreservation and either high cryoprotectant concentrations coupled with very fast cooling rate (vitrification) or lower cryoprotectant concentration and slow cooling rate (slow freezing) are utilized. In some cases researchers showed that, when using liquid nitrogen slush and techniques that allow ultra-rapid cooling rates of up to 250,000°C/min, the concentration of permeating cryoprotectants can be decreased to about the same level as used in slow freezing (Lee et al. 2010; Risco et al. 2007). Despite many advances in the field of cryopreservation, oocyte (ovulated, mature or immature) cryopreservation is far more difficult than embryo cryopreservation and is thus still not considered a fully matured procedure. The first reports on successful embryo cryopreservation were published in 1971-2 (Whittingham 1971; Whittingham et al. 1972; Wilmut 1972) and the modification to cooling rate that came a few years later (Willadsen et al. 1978; Willadsen et al. 1976), took place more than two decades after Polge and co-workers reported their success in freezing spermatozoa (Polge et al. 1949). When considered from conservation standpoint, embryo freezing has the advantage of preserving the entire genetic complement of both parents. Naturally, both male and female embryos should be stored to ensure representation of both sex chromosomes and a wide genetic diversity. Similarly, preservation of both male and female somatic cells for nuclear transfer can potentially achieve the same goal. While the "sex" of somatic cells is known since the sex of the donor is easily verified, for embryos a complicated, invasive and time-consuming procedure should be performed. The alternative would be to bank large number of embryos, assuming that some will be male embryos and some female. Cryobanking of embryos can thus help establishing founder population with the aim of eventual reintroduction into the wild (Ptak et al. 2002). However, evolution has made each species unique in many respects, one of which is the development of a highly specialized reproductive adaptation (Allen 2010; Wildt and Wemmer 1999). Thus, what may work for one species, does not necessarily work for another. While thousands on thousands of offspring were born following the transfer of frozen-thawed embryos in humans, cattle, sheep and mice, success is very limited in many other, even closely related species. So, obviously the best option is to test and make the necessary adjustments to protocols using the embryos of the target species. In wild, and especially endangered species, this is often almost impossible and the opportunity to collect oocytes or embryos is very rare for most species. To overcome this limitation, researchers find it imperative to use laboratory, farm or companion animals as models during the process of developing the necessary reproductive techniques associated with embryo cryopreservation. These techniques include such procedures as oocyte retrieval (with or without hormonal intervention; ante- or post-mortem), *in vitro* maturation, *in vitro* fertilization, *in vitro* culture and embryo transfer to recipients at the appropriate time. In some instances suitable model species were found. For example studies on the domestic cat helped develop various technologies, which were later used in non-domestic felids (Dresser et al. 1988; Pope 2000; Pope et al. 1994), or cattle served as a model for other ungulates (Dixon et al. 1991; Loskutoff et al. 1995). Unfortunately, for many species (e.g. elephant, rhino) no suitable model can be located and studies should be conducted with the limited available resources while relying on the already available knowledge from studies on closely related species (Hermes et al. 2009b). As will be discussed in the following text, cryopreservation of embryos in many mammalian species shows some, though very limited, success. The situation is much less advanced in all other vertebrates (fish, birds, reptiles and amphibians) where noticeably less efforts have been invested and the challenges are often considerably more complex. In comparison to mammals, embryos in all these classes are usually much larger in volume, with large amount of yolk and multiple membranes showing varying permeability to water and cryoprotectants. All these make embryos in these classes highly susceptible to chilling injury and, with the currently available knowledge and techniques, make their cryopreservation extremely complicated and often practically impossible (Pearl and Arav 2000).

Two basic embryo cryopreservation techniques currently rule the field. The first to be developed is the slow freezing technique (Whittingham et al. 1972; Willadsen et al. 1978; Willadsen et al. 1976; Wilmut 1972). Following this technique, embryos are gradually exposed to relatively low concentration of cryoprotectants. These are usually glycerol or Me_2SO in the range of 1.35 to 1.5M, which are added to the embryo culture medium. The embryos are then loaded in a small volume into a straw, cooled down to about -5 to -7°C where they are kept for several minutes to equilibrate. After equilibration, the solution is seeded to initiate extracellular freezing and then cooled slowly, at about 0.3 to 0.5°C/min to anywhere between -30°C and -65°C. Once at the desired temperature the straws with the embryos are plunged into liquid nitrogen for storage. When following this procedure, seeding of the extracellular solution and the very slow cooling rate ensure that freezing will take place only outside the embryo, resulting in the outward movement of osmotically-active water from the embryos and their gradual dehydration. The second technique is called embryo vitrification and it follows the same principles as described earlier for oocytes. The most common vitrification technique is simply to plunge the straw with the embryo into liquid nitrogen. While this is a simple method, the contact between the straw and the liquid nitrogen bring the latter to its boiling point. Bubbles and nitrogen gas thus form an insulation layer around the straw, which impedes efficient heat transfer (Cowley et al. 1961). To minimize this effect, faster cooling rates such as the liquid nitrogen slush (Arav et al. 2000b) using devices such as the VitMaster® (Figure 9) are employed. To increase the probability of vitrification by increasing the viscosity of the medium in which the embryos are suspended, very high concentrations of cryoprotectants, most commonly glycerol, Me_2SO, ethylene or propylene glycol, are used in combination with various sugars. The concentrations used, usually in the range of 5M or more, are a compromise between the very high concentrations required to ensure vitrification and the toxicity of such high concentrations to the embryos. To avoid osmotic shock, the embryos are exposed to the vitrification solution in a step-wise method going gradually up in the cryoprotectant concentration. The third component in the 'probability of vitrification' formula is the volume – the smaller the volume, the lower the chance for nucleation and the better the heat transfer and thus the higher the probability of vitrification (Arav 1992; Arav et al. 2002a; Yavin and Arav 2007). A large number of techniques have been developed to reduce the sample volume. These can be divided into two general groups – those that use surface carriers and those that use tubing carriers. The surface carriers techniques include electron microscope grids (Martino et al. 1996; Steponkus et al. 1990), Minimum Drop Size (MDS; Arav and Zeron 1997; Yavin and Arav 2001), Cryotop (Hamawaki et al. 1999; Kuwayama and Kato 2000), Cyroloop (Lane et al. 1999a; Lane et al. 1999b), hemi-straw (Hamawaki et al. 1999; Kuwayama and Kato 2000; Vanderzwalmen et al. 2000), solid surface vitrification (Dinnyes et al. 2000), nylon mesh (Matsumoto et al. 2001), Cryoleaf^{TM} (Chian et al. 2005), direct cover vitrification (Chen et al. 2006c), fibre plug (Muthukumar et al. 2008), vitrification spatula (Tsang and Chow 2009), Cryo-E (Petyim et al. 2009), plastic blade (Sugiyama et al. 2010) and Vitri-inga (Almodin et al. 2010). The tubing carrier techniques include the plastic straw (Rall and Fahy 1985), Open Pulled Straw (OPS; Vajta et al. 1997; Vajta et al. 1998), minimum volume cooling (MVC; Hamawaki et al. 1999), Closed Pulled Straw (CPS; Chen et al. 2001), flexipet denuding pipette (Liebermann et al. 2002), superfine OPS (SOPS; Isachenko et al. 2003b), CryoTip$^{®}$ (Kuwayama et al. 2005), high-security vitrification device (HSV; Camus et al. 2006), quartz capillaries (Risco et al. 2007), pipette tip (Sun et al. 2008), Sealed Pulled Straw (SPS; Yavin et al. 2009), Cryopette$^{®}$ (Portmann et al. 2010), Rapid-i^{TM} (Larman and Gardner 2010) and JY Straw (R. C. Chian, personal communication). Each of these two groups has its specific advantages. In the surface methods, the size of the drop (~0.1 µL) can be controlled, high cooling rate can be achieved because these systems are open, and high warming rates are achieved by direct exposure to the warming solution. The tubing systems have the advantage of achieving high cooling rates in closed systems, thus making them safer and easier to handle.

Decreasing the vitrified volume and increasing the cooling rate allows a moderate decrease in the cryoprotectant concentration so as to minimize its toxic and osmotic hazardous effects (Criado et al. 2010; Lee et al. 2010; Yavin et al. 2009). Unlike the slow cooling rate cryopreservation method, which requires sophisticated equipment to control the cooling rate, vitrification can be done relatively cheap and even under field conditions with no need for special equipment, making it a good alternative for the use in various settings often encountered with wildlife species – zoos, poorly equipped locations and field work in remote sites. However, performing vitrification, and in particular loading the sample properly into or onto the carrier, does require much experience to be done properly. The drawback of both controlled-rate freezing and conventional vitrification is that non-equilibrium freezing is taking place.

The result is elevated risk of recrystalization or devitrification unless very fast cooling and warming rates are used. To avoid this risk, an alternative method, named 'equilibrium vitrification' was proposed (Jin et al. 2010). To do this, slow cooling rate that will allow sufficient dehydration of the germplasm is required. In the slow freezing technique this can be achieved if slow cooling progress to temperatures as low as -80°C before plunging the sample into liquid nitrogen. To achieve near equilibrium freezing for vitrification, the paradoxical slow cooling rate vitrification was used. Following this technique, the straw with the sample is placed over liquid nitrogen vapor (about -150°C) for three minutes before being plunged into the liquid nitrogen. A cooling rate of ~300°C/min are thus achieved for cooling the germplasm down to below the cytoplasm glass transition temperature (about -130°C). High proportion of embryos vitrified by this technique was able to survive for days at relatively high temperature (-80°C). Yet another technique for embryo vitrification that has been proposed recently is using super-cooled air for vitrification (Larman and Gardner 2010; Larman and Gardner 2011). Following this technique, a straw is first inserted into the liquid nitrogen to cool the air inside and then the device with the embryo (mouse embryos in this case) is inserted into the straw.

Besides the production and preservation of embryos for ET, embryos can be a source for primordial germ cells (PGC), which, as was shown in the zebrafish, can be vitrified, warmed and then transplanted into sterilized recipient blastulae to differentiate into males and females that produced gametes carrying the genetic material of the transplanted PGC donor (Higaki et al. 2010). Going even earlier in the development timeline, embryos can be a source for stem cells. Embryonic stem cells, being pluripotent, can differentiate in vivo or in vitro into germ cells so they, too, can be considered an optional venue. For example an embryonic stem cell line was derived form human blastocysts (Thomson et al. 1998). Or, in a study on mice, embryonic stem cells transplanted into recipient mice were able to form testicular tissue structures and direct spermatogenesis (Toyooka et al. 2003). These cells, which can be isolated from embryos, can also be cryopreserved (Thomson et al. 1998; Toyooka et al. 2003) or vitrified (Reubinoff et al. 2001). Embryonic stem cells can also be derived from isolated blastomeres, and blastomers can also be cryopreserved individually by inserting them into emptied zona pellucida and then vitrifying them (Escriba et al. 2010) in a similar manner to the cryopreservation of single sperm described above.

In the following sections, advances in wildlife embryo cryopreservation in the various classes (mammals, birds, fish, amphibians and reptiles) will be reviewed. Mammals will occupy the bulk of the discussion, as they are the only class in which some success has been achieved thus far (Table 2).

Table 2. List of species in which attempts at embryo cryopreservation have been reported

Species	Scientific name	Pregnancy	Year	Reference
Primates				
Human	Homo sapiens	Yes	1983	(Trounson and Mohr 1983; Zeilmaker et al. 1984)
Baboon	Papio sp.	Yes	1984	(Pope et al. 1984)
Marmoset monkey	Callithrix jacchus	Yes	1986	(Hearn and Summers 1986; Summers et al. 1987)
Cynomolgus monkey	Macaca fascicularis	Yes	1986	(Balmaceda et al. 1986; Curnow et al. 2002)
Rhesus macaque	Macaca mulatta	Yes	1989	(Wolf et al. 1989; Yeoman et al. 2001)
Hybrid macaque (pig-tailed & lion-tailed)	Macaca nemestrina & M. silenus	Yes	1992	(Cranfield et al. 1992)
Western lowland gorilla	Gorilla gorilla gorilla	No	1997	(Pope et al. 1997a)
Ungulates				
Bovine	Bos taurus	Yes	1973	(Willadsen et al. 1978; Wilmut and Rowson 1973)
Sheep	Ovis aries	Yes	1974	(Willadsen et al. 1974; Willadsen et al. 1976)
Goat	Capra aegagrus	Yes	1976	(Bilton and Moore 1976)
Horse	Equus caballus	Yes	1982	(Slade et al. 1985; Yamamoto et al. 1982)
African eland antelope	Taurotragus oryx	Yes	1983	(Dresser et al. 1984; Kramer et al. 1983), cited in (Schiewe 1991a)
Arabian Oryx	Oryx leucoryx	No	1983	(Durrant 1983)
Gaur	Bos gaurus	No	1984	(Armstrong et al. 1995; Stover and Evans 1984)
Bongo	Tragelphus euryceros	No	1985	(Dresser et al. 1985)
Scimitar-horned Oryx	Oryx dammah	No	1986	(Schiewe et al. 1991b; Wildt et al. 1986)

Species	Scientific name	Pregnancy	Year	Reference
Swine	Sus domestica	Yes	1989	(Hayashi et al. 1989)
Red deer	Cervus elaphus	Yes	1991	(Dixon et al. 1991)
Suni Antelope	Neotragus moschatus zuluensis	No	1991	Cited in (Schiewe 1991a)
Wapiti	Cervus canadensis	Yes	1991	Cited in (Rall 2001)
Water buffalo	Bubalis bubalis	Yes	1993	(Kasiraj et al. 1993)
Fallow deer	Dama dama	Yes	1994	(Morrow et al. 1994)
Domestic donkey	Equus acinus	Not reported	1997	(Vendramini et al. 1997)
Dromedary camel	Camelus dromedarius	Yes	1999	(Nowshari et al. 2005; Skidmore and Loskutoff 1999)
European mouflon	Ovis orientalis musimon	Yes	2002	(Ptak et al. 2002)
Llama	Lama glama	Yes	2002	(Aller et al. 2002; Lattanzi et al. 2002)
Wood bison	Bison bison athabascae	No	2007	(Thundathil et al. 2007)
Sika deer	Cervus nippon nippon	Yes	2008	(Locatelli et al. 2008)
Carnivores				
Domestic cat	Felis catus	Yes	1988	(Dresser et al. 1988)
African wildcat	Felis silvestris	Yes	2000	(Pope et al. 2000)
Siberian Tiger	Panthera tigris altaica	No	2000	(Crichton et al. 2000; Crichton et al. 2003)
Blue fox	Alopex lagopus	No	2000	Cited in (Farstad 2000b)

Table 2. Continued

Species	Scientific name	Pregnancy	Year	Reference
Ocelot	Leopardus pardalis	Yes	2001	(Swanson 2001; Swanson 2003)
Tigrina	Leopardus tigrinus	No	2002	(Swanson et al. 2002)
Bobcat	Lynx rufus	No	2002	(Miller et al. 2002)
European Polecat	Mustela putorius	Yes	2003	(Lindeberg et al. 2003; Piltti et al. 2004)
Caracal	Felis caracal or Caracal caracal	Yes	2003	(Pope et al. 2006), cited in (Swanson 2003)
Geoffroy's cat	Felis geoffroyi	Yes	2004	(Swanson and Brown 2004)
Serval	Leptailurus serval	No	2005	(Pope et al. 2005)
Dog	Canis lupus familiaris	Yes	2009	(Suzuki et al. 2009)
Clouded leopard	Neofelis nebulosa	No	2009	(Pope et al. 2009)
Glires				
Mouse	Mus musculus	Yes	1972	(Whittingham 1971; Whittingham et al. 1972; Wilmut 1972)
European rabbit	Oryctolagus cuniculus	Yes	1974	(Bank and Maurer 1974; Popelkova et al. 2009; Whittingham and Adams 1974; Whittingham and Adams 1976)
Rat	Rattus norvegicus	Yes	1975	(Kono et al. 1988; Whittingham 1975)
Syrian hamster	Mesocricetus auratus,	Yes	1985	(Lane et al. 1999a; Ridha and Dukelow 1985)
Mongolian gerbil	Moriones unguieulatus	Yes	1999	(Mochida et al. 1999)

Species	Scientific name	Pregnancy	Year	Reference
Marsupials				
Fat-tailed dunnart	Sminthopsis crassicaudata	No	1994	(Breed et al. 1994)
Others (fish)				
Zebrafish	Danio rerio		1996	(Zhang and Rawson 1996)
Turbot	Psetta maxima		2003	(Robles et al. 2003)
Flounder	Paralichthys olivaceus		2005	(Chen and Tian 2005)
Gilthead seabream	Sparus aurata		2006	(Cabrita et al. 2006)
Red seabream	Pagrus major		2007	(Ding et al. 2007)

Studies in which pregnancy and/or live births have been recorded following the transfer of frozen-thawed embryos are noted. This is irrelevant for the fishes. When both slow freezing and vitrification were conducted, reference was provided for both.

Figure 10. Hamadryas baboon (*Papio hamadryas*), a member of the Papio species, which was the first non-human primate in which embryo cryopreservation was demonstrated. Photo by Eyal Bartove ©.

MAMMALS

Non-Human Primates

Probably hundreds of thousands (if not millions) of cryopreserved human embryos have been successfully transferred since the first report on a pregnancy resulting from the transfer of a frozen-thawed embryo (Trounson and Mohr 1983) and many more are stored under liquid nitrogen in many holding facilities around the world. Yet, despite the close relatedness of non-human primates to us, and the fact that they are often used as laboratory models for humans in many studies, progress in primate embryo cryopreservation has been very limited (Mazur et al. 2008) and reports are scarce but with promising results. It seems that in the field of embryo cryopreservation, humans act as model for other primates rather than the other way around. The first report on the birth of a non-human primate (baboon; Papio sp.; Figure 10) following transfer of frozen-thawed embryo came in 1984 (Pope et al.), about a year after the reported pregnancy resulting from frozen-thawed embryo in humans (Trounson and Mohr 1983). Six *in vivo-*

produced embryos were retrieved and frozen using glycerol as cryoprotectant. All six embryos survived the freeze-thaw procedure and resulted in two pregnancies (33.3%) after being transferred to six recipients. Similar reports on the cryopreservation of *in vivo*-produced embryos in the marmoset monkey (*Callithrix jacchus*) showed even higher pregnancy rates (Hearn and Summers 1986; Summers et al. 1987). In one of these studies, for example, 7/10 frozen-thawed 4 to 10-cell embryos and 5/9 frozen-thawed morulae resulted in pregnancies (Summers et al. 1987). Five pregnancies of the first and 4 of the latter were carried to term, resulting in 6 babies in each group. These authors noted that 1.5 M Me_2SO was superior to 1.0 M glycerol; the latter causing severe osmotic damage.

Many studies have indicated that *in vivo*-produced embryos survive cryopreservation considerably better than those produced *in vitro* (Hasler et al. 1995; Fair et al. 2001; Martinez et al. 2006), which is why researchers in non-human primates and, as will be discussed later, in other mammals often started their embryo cryopreservation research using *in vivo*-produced embryos. However, to collect *in vivo*-produced embryos from wildlife species requires multiple stress-inducing capturing/immobilizations, something that should be avoided when possible. The development of *in vitro* embryo production followed by cryopreservation is therefore desirable. In the cynomolgus macaques (*Macaca fascicularis*) pregnancy was reported following the transfer of *in vitro* produced, frozen-thawed embryos (Balmaceda et al. 1986). This report followed a previous one by the same group demonstrating success in IVF followed by embryo transfer without cryopreservation (Balmaceda et al. 1984). In the freezing experiment, 56 cynomolgus macaque embryos were frozen at the 4 to 8-cell stage using 1.5 M Me_2SO as cryoprotectant and the controlled-rate freezing technique. After thawing, 39 embryos (70%) were still viable, 25 of which were transferred to 9 synchronized recipients 24 to 48h after ovulation, resulting in 3 pregnancies. Not too long later, another group reported on pregnancy from frozen-thawed transferred embryo in the rhesus macaque (*Macaca mulatta*) (Wolf et al. 1989). Superovulation in this study was achieved by the use of hormonal stimulation, to which 74% (17/27) of the females responded. On average, 18 oocytes were retrieved from each responding female, 63% of them were judged as mature. Following collection, the oocytes were inseminated *in vitro* and the resultant embryos were frozen at the 3 to 6-cell stage, using a 1,2-propanediol-based freezing protocol originally developed for humans. Embryo post-thaw survival was optimal (100%; 11/11) and after transferring two embryos to each of three recipients during the early luteal phase of spontaneous menstrual cycles one pregnancy was achieved and was carried to term. The same group also attempted *in vitro* maturation (IVM) of oocytes prior to IVF, freezing and transfer (Lanzendorf et al. 1990). Embryos collected at the germinal vesicle (GV) stage did not fertilize and fertilization rate of oocytes collected at the metaphase I (MI) stage was low (32%), even if these were matured *in vitro* to the metaphase II (MII) stage. Fertilization rate of embryos collected at the MII stage was high (93%) and 8 embryos frozen and transferred at the 2 to 6-cell stage to four recipients (2 each) resulted in 3 pregnancies and the delivery of 3 twins (75%). In the Western lowland gorilla (*Gorilla gorilla gorilla*) the associated techniques (IVM, IVF, IVC) have been adapted successfully from humans (Pope et al. 1997a). Of 8 embryos at the 2-cell stage produced *in vitro*, three were transferred without prior cryopreservation to a single female leading to a pregnancy and birth of a female infant. The other five embryos were cryopreserved with 1.5 M 1,2-propanediol as cryoprotectant. Regrettably, freezing outcome was not reported. When working with animals on the brink of extinction, or with actually extinct animals, the availability of both males and females to provide their respective gametes is not always guarantied. When no other option remains, one may eventually resort to the creation of hybridization between the available gametes and those of closely related species. Exactly this was attempted by fertilizing *in vitro* oocytes from the non-endangered pig-tailed macaque (*Macaca nemestrina*) with sperm from the endangered lion-tailed macaque (*M. silenus*) (Cranfield et al. 1992). The resulting embryos were frozen in 1.5 M 1,2-proanediol and 0.2 M sucrose. Nine embryos were transferred to synchronized pig-tailed macaque females, one of which became pregnant and delivered a male infant. Following this route, and conducting enough backcrossing, can eventually lead to almost pure population of the target species.

As was discussed above, vitrification (Rall and Fahy 1985) is a good alternative to controlled-rate freezing and at times even superior to it. Following the lead of human and laboratory and farm animals' embryo cryopreservation, the use of vitrification was attempted and compared to the controlled-rate freezing method in non-human primates as well, with conflicting results (Curnow et al. 2002; Yeoman et al. 2001). In the cynomolgus monkey, early-stage (2 to 8 cells) embryos were vitrified using the open pulled straw (OPS) technique in comparison to the controlled-rate freezing method (Curnow et al. 2002). The vitrification solution contained 4.03 M ethylene glycol (EG) with either 5.0 M Ficoll or 5.2 M dextran. Vitrification proved to be inferior to controlled-rate freezing, regardless of the cryoprotectant combination, in cell survival rate (18 to 29% vs. 82%), embryo survival (26 to 32% vs. 90%) and cleavage rate (29 to 38% vs. 83%). In a different study, rhesus monkey blastocysts were vitrified by the Cryoloop technique which was compared to controlled-rate freezing (Yeoman et al. 2001). The cryopreserved blastocysts were produced *in vitro* through intracytoplasmic sperm injection (ICSI) to mature oocytes, followed by *in vitro* culture. Two different cryoprotectant combinations were tested for the vitrification solution – 2.8 M Me_2SO with 3.6 M EG (combination A) or 3.4 M glycerol with 4.5 M EG (combination B). Similar results were achieved when blastocysts were cryopreserved by the controlled-rate freezing technique [8/22 (36.4%) embryos survived and 1/22 (4.5%) hatched following co-culture] or by vitrification with cryoprotectants combination A [6/16 (37.5%) embryos survived and 1/16 (6.3%) hatched]. In comparison, using vitrification with cryoprotectant combination B, 28/33 (84.8%) of the blastocysts survived and 23/33 (69.7%) hatched. This last study not only achieved high embryonic survival using the vitrification technique, it has also demonstrated the suitability of this technique to overcome the problem of advanced-stage embryo preservation. Six blastocysts vitrified with cryoprotectant combination B and transferred after warming to three recipients (2 each) resulted in a twin pregnancy, which was carried to term.

Ungulates

Like in humans among the primates, embryo cryopreservation in the cattle industry is booming and has long reached a commercial level, enabling the movement of embryos between remote locations both locally and on the international market. For example, based on data published by the International Embryo Transfer Society (IETS), 324,591 frozen-thawed bovine embryos were transferred during 2008 worldwide (Thibier 2009). The real number is probably higher as not all embryo transfers are reported to IETS. To a much lesser extant, embryo transfer is found in sheep, goats, and horses and none in the pig during that year [although in previous years some embryo freezing did take place in pigs as well (e.g. Thibier 2006)]. Situation is dramatically less developed in other ungulates, although some advances have been made in recent years, primarily in ungulate species of commercial value such as camels, llama and red deer. Statements in reviews on assisted reproductive technologies in non-domestic ungulates from only a decade ago were to the effect that by that time only one successful embryo cryopreservation has been reported (Holt 2001). To start with, non-domestic ungulates usually do not show discernable signs of estrous and their receptive period is fairly short. This requires a thorough understanding of the estrus cycle endocrine activity and methods for it's monitoring in each species under study. It also requires the development of species-specific hormonal administration for ovarian stimulation. As in all other wildlife species, one should always keep in mind that what works for one species not necessarily would also work for another, even among closely related species. For instance, the bovine IVC protocol works well for the water buffalo (*Bubalis bubalis*) but when this protocol was used for the African buffalo (*Syncerus caffer*; Figure 11), embryos did not develop beyond the morula stage (Loskutoff et al. 1995). Whereas hormonal monitoring can be achieved non-invasively through fecal or urine analysis (Brown et al. 1996; Schwarzenberger et al. 1993), hormonal administration requires stress-afflicting activities such as repeated darting, general anesthesia or movement restriction by chute. Few animals are trained for these activities and chute technology saw some improvements over the years, however for the very vast majority, these activities are very stressful and holding institutions will often tend to avoid them when possible, or at least restrict them to the minimum. Thus, progress in this field among the non-domestic ungulates has been slow. Even though embryo transfer has produced live births in a number of non-domestic ungulate species, efficiency in *in vitro* technologies (IVM, IVF, IVC) has been low. For example, in the Kudu (*Tragelaphus* sp.), of 397 oocytes collected 79 zygotes cleaved, yet only 2 blastocysts were achieved (0.5%) (Loskutoff et al. 1995). Another example is the Mohor gazelle (*Gazella dama mhorr*) in which embryos produced by IVF with frozen-thawed semen did not develop beyond the 6 to 8-cell stage (Berlinguer et al. 2008). These and many other similar studies suggest that before reaching a stage at which embryo cryopreservation is a technology worthwhile pursuing for non-domestic ungulates, all other associated skills should reach a sufficient level of maturation to support it. Having said that, some attempts, often heavily relying on domestic species as model animals, were reported and, at times, with some measure of success.

During the efforts to develop the necessary assisted reproductive technologies for the European mouflon (*Ovis orientalis musimon*), a wild sheep threatened by extinction, the domestic sheep was used as a model. One group used a vitrification protocol developed for sheep to vitrify *in vivo*-produced mouflon embryos (Naitana et al. 2000; Naitana et al. 1997). The vitrification solution contained 3.4 M glycerol and 4.6 M EG as cryoprotectants. Of the five vitrified blastocysts, 4 survived and were transferred to 4 synchronized domestic sheep ewes, two of which became pregnant and one pregnancy was carried to term. In another study, using 25% glycerol (v/v) and 25% EG (v/v) as cryoprotectants (which results in somewhat different molar values from the above), *in vitro* produced mouflon blastocysts were vitrified (Ptak et al. 2002). Of the 23 cryopreserved embryos, 20 were transferred to domestic sheep foster mothers (2 embryos each). At 40 days, 7 of the sheep were pregnant but only 3 carried the pregnancy to term, delivering 4 normal mouflon offspring. The knowledge obtained from freezing bovine and domestic sheep embryos was used when researchers attempted to preserve the scimitar-horned Oryx (*Oryx dmmah*) embryos, however here success was limited to *in vitro* culture at best. In one study, *in vivo* produced scimitar-horned Oryx embryos were frozen in PG or glycerol but no specific results were reported (Wildt et al. 1986). In another, more comprehensive study performed on scimitar-horned Oryx, thirty late morula- to blastocyst-stage embryos were frozen in cryoprotectant containing 1.5 M Me_2SO, 1.5 M glycerol, or 1.5 M PG, 10 embryos in each (Schiewe et al. 1991b). Survival was higher in the Me_2SO and glycerol groups than in the PG group. A third of all embryos experienced damages to the zona pellucida. The surviving embryos were either *in vitro* cultured or transferred. Although the majority of *in vitro*-cultured embryos (67%) developed into hatched blastocysts after 48 h, no pregnancies were established following nonsurgical (n = 8) or laparoscopic (n = 1) transfer of the remaining transferable embryos. Another Oryx species in which an attempt to freeze embryos was made is the Arabian Oryx (*Oryx leucoryx*). Morula-stage *in vivo*-produced embryos were collected, one of which was frozen in 1.5 M Me_2SO. After thawing, the embryo was rated as having a good quality grade. It was surgically transferred to a scimitar-horned Oryx foster female but failed to produce a pregnancy (Durrant 1983). Another reported failed attempt was concerning the efforts to cryopreserve the suni antelope (*Neotragus moschatus zuluensis*) 8 cell-stage embryos in 1.4 M glycerol (N. Loskutoff, personal communication cited in Schiewe 1991a). Eighteen embryos were frozen. Of them, 9 completely degenerated after thawing and the other nine were graded as partially damaged. Despite the fact that all 9 embryos exhibited partial blastomere degradation, they were transferred by laparoscopy. No pregnancies were achieved. Attempts were also exerted to freeze *in vivo*-produced embryos of the African eland antelope (*Taurotragus oryx*) and the bongo (*Tragelaphus euryceros*) using 1.35 M glycerol as cryoprotectant and the controlled-rate freezing technique. Post-thaw evaluations indicated that 6/7 eland (Dresser et al. 1984) and 6/7 bongo (Dresser et al. 1985) embryos were considered viable and of good enough quality for transfer. Still, only one pregnancy (eland) was carried to term and even this resulted in a stillborn due to dystocia. These failed attempts were followed by subsequent transfer trials that resulted in the birth of a live eland offspring (B.L. Dressen, personal communication cited in Schiewe 1991a).

The red deer (*Cervus elaphus*) is an animal of commercial value in various parts of the world, and farming for hunting and game meat makes assisted reproduction technologies worth pursuing for this industry. *In vivo*-produced red deer embryos were frozen in 1.4 M glycerol by the controlled-rate freezing technique. The frozen embryos were shipped to another country (from New Zealand to Australia) where they were thawed and transferred. Pregnancy rate ranged between 50 and 72% in different farms with an average of 61.2% (153/247) (Dixon et al. 1991). In another study on red deer embryos, slow freezing was compared to vitrification by the OPS technique and fresh embryos as control (Soler et al. 2007). Pregnancy rates were 64.3% (18/28), 53.3% (8/15) and 70.0% (7/10) for fresh, vitrified and slowly frozen of embryos, respectively. These studies suggest that embryo cryopreservation can achieve good results and give hope that the knowledge from the red deer can be applied with similar success to other, closely related, species. The knowledge was actually used to freeze embryos from the fallow deer (*Dama dama*; Figure 12) (Morrow et al. 1994). *In vivo*-produced embryos, following chemically-induced superovulation and AI, were surgically collected from fallow deer and transferred either fresh or following cryopreservation. The cryopreservation protocol used was the one developed and successfully used for red deer (Dixon et al. 1991), however success in the fallow deer was lower than in the red deer. Pregnancy rate of frozen-thawed embryos was half that of fresh (26% vs. 53%) and the overall efficiency of the program was very low (0.9 to 1.0 surrogate pregnancy per superovulated donor). Another deer species in which work has been carried out in an attempt to freeze embryos is the sika deer (*Cervus nippon nippon*). Here, too, the protocol developed for the red deer (Dixon et al. 1991) was used. Of 142 oocytes collected subsequent to chemical synchronization, 57 (40.1%) cleaved following IVF and 14 of them reached the blastocyst stage. These embryos were frozen by the controlled-rate freezing technique and were later transferred (2 per recipient) to synchronized red deer surrogate hinds. One of the seven recipients was confirmed pregnant and delivered a healthy young sika deer fawn after 224 days of pregnancy (Locatelli et al. 2008).

Figure 11. When bovine *in vitro* embryo culture protocol was applied to its closely related African buffalo (*Syncerus caffer*), embryos did not develop beyond the morula-stage. Photo by Eyal Bartove ©.

Figure 12. The Persian fallow deer (*Dama dama mesopotamica*). Application of the red deer embryo freezing protocol to this species showed a considerably lower efficiency rate. Photo by Eyal Bartove ©.

The domestic cow has acted as a model for other members of the Bovinae subfamily. The gaur (*Bos gaurus*), a member of this subfamily living in the forested areas of South and South East Asia is classified in the IUCN red list as vulnerable. Following protocols developed for the cow, 9 *in vitro* produced blastocysts were cryopreserved in 1.4 M glycerol. One embryo was transferred to a domestic cow, which was confirmed pregnant on day 135 after the transfer (Armstrong et al. 1995) but it is not known if that pregnancy was carried to term. The cryopreservation of gaur embryos at the blastocyst stage was reported more than a decade earlier (Stover and Evans 1984), however that report did not elaborate on the freezing protocol nor was any information as to the outcome of the procedure provided. Another member of this subfamily is the wood bison (*Bison bison athabascae*), a sub species of the North American bison. Embryos were produced *in vitro* and then vitrified, using IVM, IVF, IVC and vitrification protocols developed for cattle (Thundathil et al. 2007). However, a protocol that works very well for cattle, gave fairly poor results in the wood bison. Only 6.9% (11/160) of the embryos have reached the blastocyst stage. Embryos in the morula stage (n=27) and blastocyst stage (n=6) were vitrified in 7.0 M EG, 0.5 M galactose and 18% Ficoll. Regrettably the researchers failed to report on the evaluation of the embryos after warming.

As for other members of the ungulate group, some but very modest success have been reported on the cryopreservation of domestic species like the horse (Barfield et al. 2009; Choi et al. 2009; Yamamoto et al. 1982) and swine (Dobrinsky et al. 2000; Nagashima et al. 1995), but very little success have been reported in other species, and the little that was reported is on species of some commercial value like camels and llama. Camelids are seasonal breeders and induced ovulators. Embryos of the dromedary camel (*Camelus dromedarius*), collected at the expanding blastocyst stage, were frozen by the controlled-rate freezing technique using 8 different cryoprotectant solutions (Skidmore and Loskutoff 1999). After thawing and 20h co-culture only 3/34 embryos survived, and even they they, although graded as expanding, were of poor quality. All three surviving embryos, which were all frozen with ethanediol as cryoprotectant, were transferred to a single synchronized female that, after three months, was confirmed, using ultrasnonography, to be carrying a single viable fetus. It was not reported if this fetus was carried to term. Vitrification was also tested as a method to preserve dromedary camel embryos. The *in vivo*-derived embryos were vitrified in 7.0 M EG. Post-thaw survival and intact morphology were high (92%) and following transfer of 45 embryos (20 during the breeding season and 25 off-season), 3 pregnancies were obtained, one of which was carried to term (Nowshari et al. 2005). Among the South American camelids, attempts have reached some level of success in the Llama (*Lama glama*), which is farmed in several regions of the world. Measurements indicated that the llama's embryos were found to be 3 to 5-fold larger than bovine embryos of the same developmental stage (Lattanzi et al. 2002). Being so much larger make these embryos more susceptible to chilling injury. The larger size also results in lower surface-to-volume ratio and thus less efficient movement of water and cryoprotectants across the membrane, making freezing more difficult. In one attempt, *in vivo* produced hatched blastocysts were either vitrified or frozen by the controlled-rate freezing technique (Lattanzi et al. 2002). Post-warming/thaw evaluation measure was re-expansion of the embryos and this happened after 24h in 64% (21/33) of the vitrified embryos and in 63% (12/19) of the embryos frozen by the controlled-rate freezing technique. In another attempt to vitrify Llama embryos, 10/40 embryos re-expanded after warming (von Baer et al. 2002). Three fresh-chilled and two vitrified-warmed embryos were transferred to synchronized recipients but only one of the fresh embryos resulted in a pregnancy. A third group, reporting at about the same time, was more successful, proving vitrification to be a suitable technique for the preservation of embryos of this member of the camelid family (Aller et al. 2002). *In vivo* produced embryos were collected non-surgically and vitrified at the expanded blastocyst stage in vitrification solution containing 20% glycerol, 20% EG, 3% poly-EG, 0.3 M sucrose and 0.375 M glucose. Eight embryos were transferred after warming to 4 recipients (2 each) and two of them became pregnant, each delivering one offspring.

Carnivores

The companion carnivores, the dog and the cat, are used in scientific research for a variety of purposes. As females of these animals are often spayed, ovaries can regularly be obtained for research on the development of the needed *in vitro* technologies. However, as said before, species are different in many ways, and so while in one much success have been achieved, in the other researchers are still struggling to get things to work. Probably the most advanced group amongst the carnivores in terms of embryo cryopreservation is that of the felids. This is most likely thanks to the extensive research conducted on the domestic cat (*Felis catus*), which was found to be a very suitable model for other felid species. The relative ease in obtaining cat oocytes by retrieving them from the ovaries of spayed queens undoubtedly contributed immensely to the progress in this field. Most, but not all feline species are seasonal breeders (Brown and Wildt 1997; Swanson and Brown 2004). As in the camelids, felines are also induced ovulators (the release of LH that leads to ovulation is induced by mating). The first report on successful IVF and IVC to the blastocyst stage in cats came in 1977 (Bowen). Eleven years later the first in-depth study of cat IVF and the first report on the birth of live kittens after embryo transfer of *in vivo*-derived embryos cryopreserved at the morula stage, were published (Dresser et al. 1988; Goodrowe et al. 1988). Cryopreservation was carried out using the controlled-rate freezing technique and 1.35 M glycerol as cryoprotectant, however success rate of embryo transfer was fairly low (14.4%, 17/118), most probably because all thawed embryos, regardless of their grade and level of damage, were transferred. Subsequently, offspring were produced after transfer of *in vitro*-derived embryos from *in vivo* and *in vitro* matured oocytes and with or without post thaw culture (Pope et al. 2002; Pope et al. 1994; Pope et al. 1997b; Wolfe and Wildt 1996; Wood and Wildt 1997).

Despite many similarities, differences between the domestic cat and other feline species still exist and transfer of knowledge is not entirely straightforward. Still, following the success in the domestic cat, maturation and fertilization of oocytes from a large number of feline species was demonstrated *in vitro* (Johnston et al. 1991). This included tiger (*Panthera tigris*), lion (*Panthera leo*), leopard (*Panthera pardus*; Figure 13B), jaguar (*Panthera onca*), snow leopard (*Panthera uncia*), puma (*Felis concolor*), cheetah (*Acinonyx jubatus*; Figure 13A), clouded leopard (*Neofelis nebulosa*), bobcat (*Lynx rufus*), serval (*Felis serval*), Geoffroy's cat (*Felis geoffroyi*), Temminck's golden cat (*Felis temmincki*), and leopard cat (*Felis bengalensis*). A total of 846 oocytes were recovered from ovaries of 35 individuals of these 13 species, 508 of which were of fair to excellent quality. Matured oocytes were achieved in all species but fertilization was not achieved in the jaguar, cheetah, clouded leopard, bobcat and Temminck's golden cat. IVF success rate, though, was very low - only 4 embryos (0.8%) cleaved – one of the leopard using homologous sperm and three of the puma using domestic cat sperm to create hybrid embryos. Some success in IVF came from other groups at about the same time, e.g. in the tiger (Donoghue et al. 1990) or the Indian desert cat (*Felis silvestris ornata*) (cited in Pope 2000). In the tiger, for example, oocyte fertilization rate was 63.4% and almost all of them (95.9%) progressed to the 2-cell stage. Eighty-six embryos were transferred to 6 recipients, one of which became pregnant and delivered 3 cubs. This success

followed a similar procedure but with no ET conducted a year earlier by the
same group in the leopard cat (*Felis bengalensis*) (Goodrowe et al. 1989).
Apart from the success in IVF of the Indian desert cat oocytes, Pope (2000)
also mentions IVF/ET in African wildcat (*Felis sylvestris lybica*) but
pregnancy here ended in stillbirths.

Figure 13. The cheetah (*Acinonyx jubatus*) (A) and leopard (*Panthera pardus*) (B) are
two of the feline wild species to which IVM, IVF and IVC protocols developed for the
domestic cat were adapted. Matured oocytes were achieved in both species, but *in vitro*
fertilization was successful only in the leopard. Photos by Eyal Bartove ©.

Following the progress in feline IVM/IVF/IVC/ET came the time for embryo cryopreservation. Some investigators used the domestic cat as a surrogate mother to frozen-thawed embryos of similar-sized wild feline species. Offspring have been produced this way in the ocelot (*Felis pardalis*) (Swanson 2001) and the African wildcat (Pope et al. 2000). Throughout the history of feline embryo cryopreservation, transfers of frozen-thawed embryos to conspecific recipients have often failed to produce live offspring. In a study on the clouded leopard, it was shown that females can be chemically stimulated by gonadotropins, embryos can be produced *in vitro* by either IVF or ICSI with cooled or frozen-thawed semen, however 24 frozen-thawed or 28 control fresh embryos surgically transferred to clouded leopard females did not result in any pregnancy (Pope et al. 2009). Similarly, oocytes collected from two serval females were *in vitro* inseminated with fresh or frozen-thawed semen for one female donor and only frozen semen was used for the other (Pope et al. 2005). Fresh and frozen semen resulted in similar cleavage frequency - 68% (54/80) and 64% (65/102), respectively. Embryos at the morula stage were frozen by the controlled-rate freezing technique but transferred thawed embryos did not result in any pregnancy. In the bobcat (*Lynx rufus*), embryos produced *in vivo* were used to test embryo transfer. One embryo was frozen by the controlled-rate freezing technique with 1.4 M glycerol as cryoprotective agent. Two fresh embryos and one frozen-thawed embryo were transferred to recipients. Although no pregnancy was obtained from the frozen-thawed embryo, one of the two fresh embryos was detected during ultrasound examination two weeks after the transfer (Miller et al. 2002). Failure, however, was not a universal phenomenon. In the ocelot, over 80 *in vitro*-produced embryos were cryopreserved for safekeeping. These represent 15 founders of the North American population of this species (Swanson 2003). Following laparoscopic transfer of frozen-thawed embryos, two pregnancies were established (Swanson 2006). The tigrina (*Leopardus tigrinus*), another South American small wildcat species, is listed on the IUCN Red List as vulnerable (http://www.iucnredlist.org/). *In vitro*-produced tigrina embryos were cryopreserved in 1.5 M EG, similar to the ocelot (Swanson et al. 2002). A total of 52 embryos were frozen but the researches did not report on post-thaw evaluation. In the caracal (*Caracal caracal* a.k.a. *Felis caracal*), from 452 recovered matured oocytes, 297 embryos were produced *in vitro*. Additional 16 embryos were produced following IVM of 83 oocytes. A total of 109 embryos were frozen using controlled-rate freezing in 1.4 M PG. Of nine recipients, 3 became pregnant and 3 kittens were delivered (Pope et al. 2006). In another report, Siberian tiger (*Panthera tigris altaica*) oocytes were collected by laparoscopy from chemically stimulated ovaries. These were fertilized *in vitro* with frozen-thawed semen and then cultured *in vitro* to the 2 to 4-cell stage. The resulting embryos were split between vitrification (n=70) and controlled-rate freezing with either PG (n=68) or EG (n=21). None of the embryos frozen by the controlled-rate freezing technique in either of the cryoprotectants survived. From the embryos vitrified in cryoprotectant containing 2.3 M Me_2SO, 3.0 M EG and 0.5 M sucrose, 46% (32/70) survived and showed some cleavage following IVC for 24h post-warming (Crichton et al. 2000; Crichton et al. 2003).

While in felids some limited success has been achieved, situation is much meager in canids. In the domestic cat all associated technologies have been mastered including chemical stimulation, oocyte retrieval, IVM, IVF, IVC, embryo cryopreservation and ET. Status is lagging far behind in the domestic dog (*Canis lupus familiaris*) and progress has been slow (Farstad 2000a; Farstad 2000b). Even today, preliminary technologies such as IVM, IVF and IVC are not fully mastered and outcome is often unpredictable, possibly because the media and conditions used for *in vitro* culture are not yet optimized for this species (Mastromonaco and King 2007; Rodrigues and Rodrigues 2006). In the vast majority of studies, dog zygotes do not progress to the advanced embryonic developmental stages – morula and blastocyst (Rodrigues and Rodrigues 2006). Embryo cryopreservation in the dog, leading to pregnancy after ET was reported only in 2007 (Abe et al.) and pup delivery following embryo cryopreservation was first reported two years later (Suzuki et al. 2009). Although attempts at cryopreservation of other canids' oocytes were reported (e.g. Boutelle et al. 2010) we were able to find only a single indication of embryo cryopreservation in a non-domestic canid – the blue fox (*Alopex lagopus*). Embryos were cryopreserved by both controlled-rate freezing and OPS vitrification and were transferred after thawing/warming to recipients. Although no live young were achieved, two implantation sites from each of the two cryopreservation methods were found (Personal communication with H. Lindeberg, cited in Farstad 2000a).

Some progress has also been reported in other carnivore families. In the Mustelidae family some species, such as the European polecat (*Mustela putorius*) or the American mink (*M. vison* or *Neovision vison*), are of commercial value, primarily in the fur industry. Other mustelids are listed as endangered or critically endangered, including the black-footed ferret (*M. nigripes*) and the European mink (*M. lutreola*). Thus, those species grown in commercial farms can act as models for developing reproductive technologies for the endangered species and for gaining the needed knowledge on specific attributes of the Mustelidae family. For example, studies were conducted on the European polecat with the aim of using it as a model for the European mink. The first successful embryo cryopreservation in this family was reported in 2003 (Lindeberg et al.). In this study, *in vivo* produced European polecat embryos were surgically recovered, frozen by controlled-rate freezing in 1.5 M EG and resulted, following surgical transfer after thawing, in 3/8 pregnancies and out of the 93 embryos transferred, a total of 9 pups were born (9.7%). In a follow-up study by the same group (Piltti et al. 2004), the first successful embryo vitrification in carnivores was achieved. European polecat *in vivo*-produced embryos were vitrified by the OPS technique in cryoprotectant solution containing 2.95 M EG, 2.32 M Me_2SO and 0.9 M sucrose. Of 98 vitrified embryos at the morula and blastocyst stages, 50 survived (51%) the cooling-warming cycle. They were transferred to 4 recipients and two of them delivered a total of 8 pups. Success rate in terms of pups per transferred embryos in the vitrification technique proved to be almost double that of the controlled rate-freezing technique (8/50; 16% vs. 9/93; 9.7%, respectively). Further improvements came with a different vitrification technique that came to be known as the "pipette tip." Using this technique, 43.6% (44/101) of the embryos survived vitrification and resulted in live births (Sun et al. 2008). The vitrified embryos following 2 or 16h incubation resulted in significantly higher success rate (71.3% and 77.4% live births, respectively) when compared to the 32h (25%) and 48h (7.8%) incubation times. Vitrification with 2 or 16h post-warming incubation did not differ from the fresh control embryos (79.2% live births).

Glires – Rodents and Lagomorphs

Embryo cryopreservation in the mouse (*Mus musculus*) was the first to be reported amongst all mammals (Whittingham 1971; Whittingham et al. 1972; Wilmut 1972), yet work in glires has solely concentrated since then on laboratory animals - mice, rats, hamster, gerbil and rabbits, which are often also used as pets. The major cryoprotectant used for freezing embryos in this group is Me_2SO, often at 1.5 M for controlled-rate freezing. In rabbits (*Oryctolagus cuniculus*), *in vivo* produced embryos cryopreserved by either controlled-rate freezing (Bank and Maurer 1974; Whittingham and Adams 1974; Whittingham and Adams 1976) or vitrification (Popelkova et al. 2009), resulted in fairly high survival rate of up to 83% and pregnancy rate (up to 70%). However, rate of young born was still relatively low, in the range of 7-

17% (Bank and Maurer 1974; Whittingham and Adams 1974; Whittingham and Adams 1976). In rats (*Rattus norvegicus*), both controlled-rate freezing and vitrification were attempted with considerably better results in the latter. In controlled-rate freezing experiments, *in vivo* produced embryos at the 2-, 4- and 8-cell stages were recovered and frozen with 3.0 M Me$_2$SO (Whittingham 1975). Post-thaw normal morphology recovery rate ranged between 65% and 68%. Percentage of embryos carried to term was low (11% for 2-cell embryos, zero for 4-cell and 9% for 8-cells. In contrast, in the vitrification study, *in vivo* produced blastocysts were vitrified in a vitrification solution containing 20.5% Me$_2$SO, 15.5% acetamide, 10% PG, and 6% Poly-EG. Of the vitrified embryos, 79% (117/149) were morphologically normal after warming. These were split between *in vitro* culture (n=48) and transfer to recipient rats (n=69). All cultured embryos progressed to expanded and hatched blastocysts and 41% (n=28) of those transferred resulted in live pups (Kono et al. 1988). The golden hamster, also known as the Syrian hamster (*Mesocricetus auratus*), is another member of this group that is in frequent use as a laboratory research subject. *In vivo* produced embryos at the 1- and 2-cell stage were flushed and vitrified by the cryoloop technique, the development of which was also the purpose of this study (Lane et al. 1999a). Of 216 vitrified hamster 2-cell embryos, 117 (54.2%) continued development to the morula/blastocyst stage after warming. Such embryos were transferred to two recipients who delivered 6 pups. In another study, *in vivo* produced embryos at the 8-cell stage were vitrified in 0.250 mL straws following a technique developed for mouse embryo vitrification (Mochida et al. 2000). Only post-warming *in vitro* development in culture was assessed and this was fairly poor, as only two out of 37 embryos have developed to the blastocyst stage. Like the hamster, the Mongolian gerbil (*Mesocricetus auratus*) is yet another rodent extensively in use for research (and as pet). *In vivo* produced embryos were vitrified in 0.250 mL straws using either one- or two-step exposure to the vitrification solution (Mochida et al. 1999). The two-step method showed superior results as 69% (38/55) of the 2-cell stage vitrified embryos developed after warming to the compacted morula stage. Following vitrification, 155 embryos that have developed to the blastocyst stage were transferred to 10 synchronized females, 3 of which became pregnant and delivered 15 pups (9.7%). In a follow-up study by the same group it was shown that embryos at later developmental stages (4-cell, morula and blastocyst) can also be vitrified and result in very high post-warming normal morphology (ranging between 87% and 100%) (Mochida et al. 2005). In this study, after transfer into recipient females, 3.3% (4/123), 1% (1/102), 5.5% (4/73), and 9.7% (15/155) developed to full-term offspring from vitrified-warmed early two-cell embryos, late two-cell embryos, morulae, and blastocysts, respectively.

Marsupials

Ovulated oocytes in marsupials are very different from those in eutherian mammals. They are much larger in size - diameter of 200 to 250 μm in the members of the dasyuridae family or 155 to 393 (mean 240.3 ± 7.14) in the phalangeridae, for example (Breed et al. 1994; Rodger et al. 1992). Ovulated oocytes have already shed off the cumulus cells (which do not form corona radiata) so they are basically nude with only the zona pelucida around them (Breed 1994). Oocytes are ovulated with a large yolk compartment that takes up much of their cytoplasm. Such yolk compartment presents considerable complications to freezing. The only report we know of on embryo cryopreservation in marsupials is about embryos of the fat-tailed dunnart (*Sminthopsis crassicaudata*), a small carnivorous dasyurid (Breed et al. 1994). *In vivo* produced embryos were harvested at the 1- to 4-cell stage and cryopreserved by either controlled-rate freezing (with 1.5 M Me_2SO as cryoprotectant) or by vitrification with two different cryoprotectant solutions (4.5 M Me_2SO plus 0.25 M sucrose or 40% EG, 30% Ficoll and 0.5 M sucrose). Post-thaw/warming evaluation comprised of only *in vitro* culture and morphology assessment. While up to 76% (31/41) of control embryos (no cryopreservation) cleaved, only 17% (2/12), 0% (0/4) and 18% (2/11) of those cryopreserved by controlled rate freezing, vitrified in Me_2SO or vitrified in EG, respectively, cleaved. Of 25 embryos frozen by controlled rate freezing, 20 (80%) appeared morphologically normal under the light microscope but many of the cells had multiple damages to intracellular organelles seen by electron microscopy.

Cetacean

Only very few studies have reported attempts at cryopreservation of marine mammals oocytes and the only ones we were able to locate were on the common minke whale. These include studies on both slow freezing (Asada et al. 2000; Asada et al. 2001a) and vitrification (Asada et al. 2001b; Fujihira et al. 2006; Iwayama et al. 2005). To date, no study reporting embryo cryopreservation in cetaceans has not been published (O'Brien and Robeck 2010b).

Non-Mammalian Vertebrates

Whereas embryo cryopreservation in mammals shows varying success from the highly successful cryopreservation of human and bovine embryos to the shambling embryo cryopreservation in canids, situation is much less advanced in all other vertebrates (fish, birds, reptiles and amphibians). This is because considerably less efforts have been invested in this direction and, even more so because of the significantly different structure of the embryo in these vertebrates, which complicates their cryopreservation. Successful and reproducible embryo cryopreservation in any member of these vertebrates is yet to be described. Research into embryo cryopreservation of non-mammalian vertebrates (NMV) has concentrated almost solely on fish (primarily the zebrafish; *Danio rerio*) and thus most of the ensuing discussion will use fishes as representatives for the NMV group. When compared to mammalian embryos, those of fish are often substantially larger, resulting in lower surface area to volume ratio. The consequence of this is relatively poor water and cryoprotectant movement across the membrane during chilling, freezing and thawing. Fish embryos contain a large yolk compartment, enclosed in the yolk syncytial layer (YSL). Behavior of the yolk during freezing defer from the behavior of other embryonic compartments, making freezing very complex. Fish (and other NMV) embryos have at least three membrane structures - YSL, plasma membrane of the developing embryo and the chorion membrane, which surrounds the periviteline space (Kalicharan et al. 1998; Rawson et al. 2000). Each of these membranes have a different permeability coefficient to water and cryoprotectants, resulting, for example, in water permeability in the range of one order of magnitude lower in fish embryos than in other animals [0.022 to 0.1 $\mu m \times min^{-1} \times atm^{-1}$ for the zebrafish (Hagedorn et al. 1997a) compared to 0.722 in drosophila (Lin et al. 1989) or 0.43 in mice (Leibo 1980)]. As if to complicate things even further, the different embryonic compartments have different water content and different osmotically inactive water content (Hagedorn et al. 1997b). The water content of the entire six-somite embryo was calculated to be 73.7%. This is divided between the yolk compartment (which occupies 61% of the embryonic volume at this stage) where total water content is 41.7% and the blastoderm (which occupies the balance 39%) where water content was calculated to be 82% (Hagedorn et al. 1997b). The value of 73.7% was later modified to 58%, suggesting that the extra water molecules are those that remained attached to the outer surface of the embryo (Hagedorn et al. 1997a). Since the chorion membrane can be removed enzymatically (by pronase) and its removal does not hinder embryonic development (Hagedorn et al. 1997c), Hagedorn et al. (1997a) suggested that the YSL was the primary barrier to the movement of crtyoprotectants and that the yolk sac reaches a lower level of cryoprotection compared to other embryonic compartments. Using magnetic resonance microscopy, the same group showed that while no cryoprotectant previously injected into the yolk was able to leave, some cryoprotectant was able to enter the blastoderm (Hagedorn et al. 1996). Attempts to solve this permeability issue by adding aquaporin 3 to the zebrafish embryonic membranes (Hagedorn et al. 2002), inserting cryoprotectants into the yolk by microinjection (Janik et al. 2000) or by exposing the embryos to ultrasound waves (Bart and Kyaw 2003; Silakes and Bart 2010) did not improve post-thaw survival. Attempts were made to test various permeating and nonpermeating cryoprotectants including methanol, Me_2SO, glycerol, 1,2-propanediol, PG, EG, trehalose, and sucrose. Embryos were shown to be very sensitive to glycerol and EG at a concentration of 1.5 M but less so to methanol, Me_2SO or PG (Hagedorn et al. 1997c). Studies have also showed that later-stage embryos were less chilling sensitive than early-stage and thus probably more suitable for freezing (Zhang and Rawson 1995). However, attempts to cryopreserve fish embryos by controlled rate freezing or vitrification generally met with lack of success (reviewed in Robles et al. 2009).

Amphibian ova have about 20-25 times larger diameter than that of human embryos (2.0 to 2.5 mm vs. ~100 μm, respectively). Very few studies on oocyte/embryo cryopreservation were done (Guenther et al. 2006; Kleinhans et al. 2006) and banking of such germplasm is still years ahead (Kouba and Vance 2009). An alternative can be the preservation of embryonic totipotent cells, which can later be injected into enucleated zygotes to generate new embryos (Uteshev et al. 2002). In their study, Uteshev et al. showed that as many as 87% of the cells survived freezing and in 3 to 5% of the cases such preserved cells were able to direct development up to the blastula stage.

EMBRYO CRYOPRESERVATION – CONCLUSION

Embryo cryopreservation has been a useful tool in the hands of embryologists, offering much hope for those concerned with animal conservation, for almost four decades now. This technology, however, is currently in use almost exclusively in laboratory animals, livestock and humans. Attempts at embryo cryopreservation in wildlife are scarce and proved successful in only a handful of mammalian species. Due to limitations in terms of basic reproduction biology understanding, scarcity of animals available for research, and shortage of funds, the study of embryo cryopreservation in wildlife show extremely slow progress. A wide variety of cryoprotectants have been tested for cryopreservation, initially by the conventional controlled rate freezing technique and, more recently, by vitrification. This latter method is especially attractive for wildlife embryo cryopreservation because it is simpler, faster, less expensive and suitable for work in remote locations, in zoos or in the field. Of course, embryo cryopreservation is not an independent technology and all other skills, including estrous detection, ovarian stimulation, oocyte and embryo retrieval, *in vitro* maturation, fertilization and culture and embryo transfer should also reach maturation to support it. While attempts to cryopreserve the embryos of all vertebrate classes should continue to enhance our understanding and overcome the many obstacles in our way, other alternatives such as oocyte, ovarian tissue or whole ovary cryopreservation should also be pursued to supplement it and enhance our ability to preserve biodiversity and genetic variability.

Chapter 4

OVARIAN TISSUE CRYOPRESERVATION

Since oocyte cryopreservation is problematic as was discussed earlier, the solutions offered are preservation of embryos or, albeit still investigational, preservation of ovarian tissue. There are two options for fertility preservation involving ovarian tissue. The first is represented by cryopreservation of ovarian cortical slices and the second, which will be discussed in the next section, is the cryopreservation of the whole ovary. Cryopreserving ovarian tissue has several advantages over oocyte or embryo cryopreservation, but it also comes with its unique complications. As was explained in the section on testicular tissue cryopreservation, tissue is a complex structure and thus presents many difficulties with respect to cryopreservation. Ovarian tissue is available any time (season, stage in cycle, age – from fetus to old to deceased), it contains large number of oocytes and, to overcome the problems associated with *in vitro* development and maturation, it can be implanted so that this can take place *in vivo* (Candy et al. 1995), or partially *in vivo*-developed oocytes can be retrieved and matured *in vitro* (Liu et al. 2001). Attempts to freeze ovarian tissue were reported already in 1951 (Smith and Parkes), only two years after Polge et al. discovered the protective effect of glycerol during freezing (Polge et al. 1949). The first live birth following ovarian tissue freezing and transplantation was reported in mice, in which the tissue was frozen to -79°C (Parrott 1960). Grafts can be transplanted to the owner of the tissue (autotransplantation), to another member of the same species (allotransplantation) or to a member of a different species (xenotransplantation). All three possibilities were successfully used to support follicular development in the grafted tissue. Although ovarian tissue grafting is usually done under the kidney's capsule because it is highly vascular, other locations like subcutaneous grafting for easy access have also been reported (Cleary et al. 2003). Subcutaneous grafting, however, produced inferior results compared to grafting under the kidney capsule. Transplantation can be to either female or male recipient (Snow et al. 2002; Weissman et al. 1999) and, interestingly, in a study on human ovarian cortex transplantation to non-obese diabetic-severe combined immune deficient (NOD-SCID) mice, more males (76.5%, 13/17) supported follicular development than females (30%, 6/20) (Weissman et al. 1999). In another study, while more xenografts were retrieved from females, the number of oocytes recovered from each xenograft was higher in those transplanted to males (Snow et al. 2002). Oocytes developed in males, however, showed reduced fertilizing ability and none of the transferred embryos resulted in implantation. One of the problems associated with ovarian tissue transplantation is the ischemic damage that leads to reduced follicular and graft survival and thus shortened ovarian function upon transplantation (Candy et al. 1997). The surviving follicles, though may grow and develop after transplantation, often contain oocytes of suboptimal quality (Kim et al. 2005). In fact, although in sheep autotransplantation of frozen-thawed ovarian cortex (Gosden et al. 1994a) and of hemi-ovaries (Salle et al. 2003) has resulted in deliveries and hormone production (Baird et al. 1999; Salle et al. 2003), duration of function was transient due to ischemia, which caused severe reduction of the total number of follicles (Liu et al. 2002). The experience in humans with ovarian cortex has also confirmed the short lifespan and thus the suboptimal quality of this approach (Kim et al. 2005). Transplanted ovarian tissue, like any transplanted tissue, carries the risk of transmitting diseases from the donor to the recipient, a risk that is greatly elevated by the need to use suppress the immune system in the recipients to reduce the risk of graft rejection. If the graft survives and is not rejected, it seems that the presence of ovaries in the recipient slows down its development (Cleary et al. 2003). The standard cryopreservation method, which seems to work for many different species, is freezing ovarian cortical tissue of the size of 1 to 2 mm^3 in cryoprotective solution containing Me$_2$SO, ethylene glycol or 1,2-propanediol. The tissue and the cryoprotective solution are equilibrated at 0°C and then again at -5 to -7°C. Seeding to initiate extracellular freezing is performed and the sample is then cooled at a slow and constant rate of 0.3-0.5°C/min till somewhere between -30° and -80°C before being plunged into liquid nitrogen for storage (for review see Paris et al. 2004). An alternative method proposed a few years ago does not requires expensive equipment and is suitable for work under field conditions (Cleary et al. 2003). Following this method, equilibration is done on ice and the sample is then placed in a passive freezing device that is placed on dry ice. Using this device, a cooling rate of about 1°C/min can be achieved. This is faster than the optimal cooling rate but still tolerable. Freezing wombat ovarian tissue following this technique resulted in 134 ± 32 intact follicles per graft compared to 214 ± 55 for the controlled-rate freezing machine. The third alternative is cryopreservation of ovarian tissue by vitrification. Toxicity of the high concentrations of cryoprotectants needed and the problem of fast enough heat dissipation throughout the tissue still pose a problem but several researchers, using a variety of vitrification carriers, demonstrated that this technique is worth pursuing. In human, for example, no difference was found between ovarian slices cryopreserved by vitrification and those cryopreserved by slow freezing in any of the *in vitro* evaluation techniques (Isachenko et al. 2009; Li et al. 2007). In both studies vitrification was achieved by dropping the ovarian slices directly into the liquid nitrogen after transferring them through a series of cryoprotectant solutions with increasing concentrations. Similar results were also achieved when the human ovarian tissue samples were packaged in 0.5 mL straws (Keros et al. 2009). Vitrification of human ovarian tissue, using three different carrier methods (0.25-mL straw, metal powder cooled in liquid nitrogen and electron microscope copper grid) was shown to be even superior to ultra-rapid freezing (liquid nitrogen vapor at about -120°C) in the levels of ROS production and apoptosis (Rahimi et al. 2003). High survival rate was also reported for bovine ovarian slices, which are often used as a model for humans, as well as in mice (Kagawa et al. 2007; Kagawa et al. 2009). In both species ovarian slices were transplanted to the respective species and resulted in the development of oocytes. In mice these oocytes were harvested, fertilized *in vitro* and resulted in the birth of live young after embryo transfer. In another study on mice ovarian tissue vitrification, slow freezing, vitrification in 0.5 mL straws and direct cover vitrification (DCV) techniques were compared (Chen et al. 2006c). The DCV technique was shown in this study to be superior to the two other cryopreservation techniques in the percentage of viable follicles, morphologically normal follicles, intact ultrastructure of primordial follicles, follicle number after transplantation, pregnancy rate (only compared to the other vitrification technique) and litter size. And, in monkeys (both cynomolgus and rhesus macaques) it was shown that follicular loss following vitrification or slow freezing was minimal when compared to fresh control (70.4 ± 4.8%, 67.3 ± 1.9% and 76.0 ± 4.1%, respectively) and that co-culture with mitotically inactivated mouse fetal fibroblasts increased survival rates in both cryopreservation methods (89.2 ± 2.1% and 90.3 ± 1.9%, respectively)

(Yeoman et al. 2005). This increase in viability was in the 30 to 50 μm follicles while some decrease in the larger follicles was noted. When testing ovarian tissue vitrification in goats, it was found that the solid surface vitrification technique produced superior *in vitro* outcome when compared to vitrification in straws (Santos et al. 2007). In mice and humans a technique named 'needle immersed vitrification' was shown to be superior to both vitrification of tissue slices by dropping them into the liquid nitrogen and to cryopreservation by slow freezing in a variety of *in vitro* and *in vivo* evaluation techniques which included transplantation and IVF of recovered oocytes (Wang et al. 2008). Others, however, did not achieve the same positive results. In a study on human, bovine and swine ovarian tissue cryopreservation, *in vitro* evaluation showed that vitrification produced inferior results when compared to the slow freezing technique (Gandolfi et al. 2006). To take things even further, researchers compared fresh mouse ovarian tissue to ovarian tissue vitrified on the fibreplug carrier (Wang et al. 2011). Preantral follicles from tissue samples of both control and experimental groups were cultured *in vitro*, oocyte maturation was chemically induced on day 12 of culture and mature oocytes were harvested. These oocytes were fertilized *in vitro*, cultured to the blastocyst stage, vitrified and later warmed and transferred to recipient mice. Although the experimental group was inferior to control in the number of matured oocytes developed in the cultured follicles and the proportion of embryos that developed to the blastocyst stage, live pups were produced from both. Cryopreservation of ovarian tissue, which is later auto-, allo- or xenografted, has been reported in a variety of species including humans (Donnez et al. 2004; Gook et al. 2003; Gook et al. 2001; Weissman et al. 1999), non-human primates - rhesus macaque (*Macaca mulatta*) (Lee et al. 2004), bovine (Herrera et al. 2002), sheep (Gosden et al. 1994b), cats (Gosden et al. 1994b; Jewgenow et al. 1997; Jewgenow and Paris 2006; Luvoni 2006), mice (Liu et al. 2001; Liu et al. 2000; Parrott 1960), rabbits (Almodin et al. 2004), common wombat (*Vombatus ursinus*) (Cleary et al. 2003; Wolvekamp et al. 2001), African elephant (*Loxodonta africana*) (Gunasena et al. 1998), tammar wallaby (*Macropus eugenii*) (Mattiske et al. 2002), and Fat-tailed dunnart (*Sminthopsis crassicaudata*) (Shaw et al. 1996). The last two are of special interest as they demonstrate that even when xenografting between as philogentically distant mammals as marsupials and mice, the graft is still supported and oocytes can develop. Whereas in most cases xenografting ovarian tissue is done into mice, xenografting of mouse ovarian tissue into rats has also been attempted and the resulting oocytes were later fertilized *in vitro*, cultured and transferred to surrogate mice, leading to the birth of healthy, fertile mice (Snow et al. 2002).

Primordial oocytes in ovarian tissue are probably less prone to cooling and freezing damages when compared to mature ones because they are smaller in size and they lack zona pellucida. Still, recovery rate is low. In cats, for example, only 10% of the follicles survived the freezing, thawing and transplantation-associated ischemia (Bosch et al. 2004). To overcome this low harvesting rate, multiple grafts are required. But, despite all difficulties, success can be achieved. For instance, a recent case report in humans described the birth of two healthy boys following ovarian cortex freezing, thawing, transplantation, controlled ovarian stimulation, oocyte retrieval and vitrification, IVF, *in vitro* embryo culture and transfer (Sánchez-Serrano et al. 2010).

Chapter 5

WHOLE OVARY CRYOPRESERVATION

Freezing large volumes, including whole organs, involves several aspects, which make any attempt at cryopreservation a challenge (Arav and Natan 2009). These difficulties include: 1) to avoid damages to the tissue, efficient heat transfer throughout the tissue should be achieved. When a thick tissue or a whole organ is involved, this is very difficult to accomplish, 2) efficient cryoprotectant penetration to all cells in the tissue should be achieved. This is challenging because of the tissue thickness and also because different cell type in the tissue have different permeability coefficients. Excessive exposure time may be damaging to some cells in the tissue due to cryoprotectant toxicity while too short exposure might leave some of the cells unprotected. Thus, the optimal time slot is to be identified, 3) supercooling (cooling below the freezing point of the solution without crystallization) may take place in some parts of the tissue. This may lead to damages from uncontrolled intra- and extracellular ice formation and faster than optimal cooling rates once freezing occurs, 4) attaining homogenous cooling rate while avoiding the excessive build-up of toxic concentrations of cryoprotectants, 5) during freezing, latent heat is released from the solution. This released heat can induce recrystalization and extend the isothermal stage, resulting in the development of a large temperature difference between the tissue/organ and the surrounding. This may lead to faster-than-optimal cooling once all latent heat has been released, 6) recrystalization may also occur during thawing because of inhomogeneous warming of the sample. Still, if these issues can be overcome, whole ovary cryopreservation presents one very important advantage over cryopreservation of ovarian slices. One of the major problems with ovarian cortical tissue cryopreservation is the ischemia the graft goes through when transplanted. This ischemia cause both graft loss and the death of a large proportion of the follicles within the tissue. Cryopreserving the whole ovary, including its vascular pedicle can ensure blood supply as soon as the organ has been transplanted (Bromer and Patrizio 2009). For the grafted ovary to become fully functional, both ovaries of the recipient should be removed (Liu et al. 2008). Grafting the ovary can be done to its natural position or to any other location in the body that may provide easy access. Of course ovary transplanted to other location can only produce oocytes that can be harvested for use *in vitro*. First whole ovary freezing, done in sheep, was reported by us a decade ago (Revel et al. 2001; Revel et al. 2004). In this study we used directional freezing technique, which is claimed to provide the solution for many of the issues involved in large volume freezing mentioned above (Arav and Natan 2009). The results of this study, albeit preliminary and gathered in one animal model, seem to confirm prolonged ovarian survival when the organ is cryopreserved as a whole and then utilized for retransplantation. Before this report, our group (Arav and Elami 2005) documented the longest ovarian function for up to 3 years after whole organ cryopreservation in sheep. Recently we extended that work by reporting, in the same animal model, the longest documented ovarian functional survival 6 years after whole organ cryopreservation, thawing and orthotropic transplantation (Arav et al. 2010).

Following the introduction of directional freezing, we were able to overcome most of the problems involved in heat and mass transfer. We will describe here, in brief, the multi-thermal gradient (MTG) device utilized for directional freezing and the protocol used for ovary preparation and slow freezing. The device is built of four temperature domains within 270 mm copper blocks (Figure 5). The test tube is advanced at a constant velocity (V) through the predetermined temperature gradient (G). The value of G is based on G= $\Delta T/d$, where ΔT is temperature differences and d is the distance between temperatures. The result is a cooling rate (B) according to the equation B = G × V. The cooling rate set to 0.3°C/min by adjusting the speed at which the tube passes through the temperature gradient. Seeding is performed at the tip of the test tube and ice interface is propagated according to the freezing point of the solution. The University of Wisconsin solution (UW) (Madison, WI, USA), used in organ transplantation, supplemented with the cryoprotectant Me_2SO, is employed for vascular perfusion.

The ovarian artery can be perfused under a microscope with cold (4°C) UW supplemented with 10% Me_2SO for 3 minutes and then inserted into a freezing tube containing the same cryoprotectant. Slow freezing is performed as follows: slow cooling to -6°C when seeding is performed. Then directional freezing commence to -14°C or to -30°C at 0.03 mm/sec or 0.01mm/sec, respectively, resulting in a cooling rate of 0.3°C/min. When reaching the target temperature, the tubes are plunged into liquid nitrogen (LN). Thawing is performed by plunging the test tube into a 68°C water bath for 20 seconds and then into a 37°C water bath for 2 minutes. Careful temperature measurements are taken to avoid heating the ovaries above 20°C during thawing.

Most other whole ovary freezing experiments reported in the scientific literature used controlled-rate freezing equipment to achieve the desired very slow (~0.1°C/min) cooling rate needed. Our first report was followed by reports on freezing ovaries of various other species such as rats (Qi et al. 2008; Wang et al. 2002), mice (Liu et al. 2008), bovine (Arav 2003), pigs (Imhof et al. 2004), rabbits (Chen et al. 2005), human (Bedaiwy et al. 2006) and other studies on sheep (Onions et al. 2009). In some of these studies, pregnancies were achieved and live young were produced. Interestingly, to date whole ovary cryopreservation and transplantation in humans has not been reported (Bromer and Patrizio 2009) despite the fact that ovarian transplantation has been in practice for several years now and whole human ovary freezing was attempted by several researchers.

Although vitrification is an attractive procedure for cryopreservation of whole ovaries, the current knowledge in cryobiology is insufficient to overcome the multiple problems involved in large volume vitrification (Fahy et al. 1990). As was discussed in the section on semen vitrification above, to reach the state of vitrification, while avoiding cryoprotectant toxicity, very high cooling rates are needed. To realize these, researchers sought methods to reduce the frozen sample volume to the very minimum possible. If vitrification attempt fails, it can be due to ice nucleation, crystal growth or fracture, three issues that are dependent on cooling velocity and temperature gradient within the frozen sample (Fahy et al. 1990). When the volume is large, as is the case in ewe, bovine or human ovaries for instance, all three damaging processes may occur. Still, attempts at whole ovary vitrification did take place and in some cases, when the ovaries were sufficiently small, were even successful. An attempt to vitrify whole sheep ovary resulted in complete loss of all follicles (Courbiere et al. 2009). On the other hand, in studies on mice, vitrification of whole ovary was successful. One study showed acceptable post-warming viability by *in vitro* evaluations (Migishima et al. 2003). When the ultrastructure of vitrified mouse ovaries was compared to control (non-cryopreserved ovaries), the quality of the oocytes and follicular cells was found to be similar. However, some of the mitochondria in the vitrified ovaries were found to be swollen (Salehnia et al. 2002). In another study, live offspring were produced when the donor mice were transgenic so that their ovaries expressed anti-freeze protein type III (Bagis et al. 2008).

Tissue preservation could be applied to endangered species is several ways. Xenografting tissue (ovaries or testis) from endangered species can be done into nude mice, as was shown, for example, in the common wombat and elephant (Cleary et al. 2003; Gunasena et al. 1998). Another option is to transfer a whole ovary, together with the preparation for anti rejection, to a recipient female (Chen et al. 2006a; Silber et al. 2007). Isolation of primordial follicles from the ovary can then be done for use for *in vitro* culture and maturation (Eppig and O'Brien 1996).

Chapter 6

FREEZE-DRYING TECHNIQUE

Storage of cryopreserved samples under liquid nitrogen is very demanding in terms of maintenance, storage space, storage equipment, costs of liquid nitrogen and the high carbon footprint of its production and transportation. An alternative that would minimize costs, storage, maintenance and carbon footprint has been gaining a foothold, primarily in the field of sperm preservation in recent years - the dry storage. Drying of cells can be achieved by either freeze-drying or convective-drying. Freeze-drying was known for hundreds of years as a method of meat and vegetable preservation among the people who lives in very high altitudes, like among the Andes mountain dwellers of South America. The Incas built thousands of storage sites and developed methods to preserve food. Freeze-drying today is accomplished with equipment developed for the space program, but the Incas achieved the same result by utilizing the cold weather and low pressure in the Andes. Potatoes were left outside at night to freeze. In the daytime, the hot sun evaporated the moisture, resulting in a freeze-dried potato pulp called chuño. The same process was used to preserve beef. In more recent times, freeze-drying has been used for preparation of pharmaceutical, viral, bacterial, fungal or yeast preparations in a dry and convenient form for transportation and storage. A significant achievement using this process took place during the Second World War when freeze-drying was used to preserve plasma and penicillin. Freeze-drying is achieved by sublimation of the ice after freezing the sample to subzero temperatures. Convective drying, on the other hand, is achieved by placing the sample in a vacuum oven at low or ambient temperatures or by applying a flow of nitrogen gas over the sample. In nature, many plants and animals can enter the state of anhydrobiosis by accumulating disaccharides such as trehalose in their cells to as much as 50% of their dry weight (Crowe et al. 1984; Westh and Ramløv 1991; Womersley and Ching 1989). Several species regularly survive in a dehydrated state even for up to ten years. Depending on the environment, they may enter this state of anhydrobiosis. While in this state, their metabolism is lowered to less than 0.01% of normal levels and their water content can drop to 1% of normal. The Tardigrades, also known as water bears, are able to enter this state and survive for years (Guidetti et al. 2002). These micrometazoans were described in the 18th century by Lazzaro Spalanzani who is also known as the first cryobiologist to show survival of frog, stallion and human sperm after chilling to 0°C in snow (Spallanzani 1776).

In order to freeze-dry cells, we first had to develop a method that would enable freezing in the absence of permeating cryoprotectant agents (CPAs) such as Me_2SO, glycerol or ethylene glycol, to name a few, and would enable the use of additives that have a high glass transition temperature (Tg) and that are solid at room temperature. Overcoming these obstacles is not simple. As Thomas A. Jennings said ''most investigators have at times overlooked the importance of the freezing process...while simple in concept, the freezing process will be shown to be perhaps the most complex and least understood step in the lyophilization process'' (Jennings 2002). The major damaging factors associated with freeze-drying of liposomes are lipid-phase transition (LPT) and fusion (Crowe et al. 1997). Cellular membrane LPT is also the mechanism underlying damage that occurs during chilling and it is the main obstacle for successful cryopreservation of many cell types, including sperm (Drobnis et al. 1993) and oocytes (Arav et al. 1996). However, cryopreservation of cells, which are more complex then liposomes, has even more damage mechanisms, on top of those to the membrane lipids. These include such mechanisms as damage to cell structure and cytoskeleton during freezing (Hosu et al. 2008; Weiss and Armstrong 1960) and damage to lysosomes and mitochondria when cells are dehydrated during the freezing process (McGann et al. 1988). Cells desiccation causes down regulation of metabolism, increases intracellular viscosity and salt concentrations, causes denaturation of proteins and creates free radicals (Allison et al. 1998; Potts 1994; Potts et al. 2005; Wang 2000). Chilling injury can be overcome by stabilizing the membrane phospholipids using disaccharides such as sucrose or trehalose (Crowe and Crowe 1991). Other approaches to decreasing the damage associated with LPT have involved changing the lipid composition of the membrane by using liposomes *in vitro* (Saragusty et al. 2005; Zeron et al. 2002b) or dietary additives *in vivo* (Zeron et al. 2002a). Altering membrane lipid composition has been shown to improve the freeze-drying of platelets (Leidy et al. 2004). The second factor is membrane fusion, which can occur when the dried cells, maintained in a fluid matrix, come into contact (Sun et al. 1996). Liposomes stored above the Tg have been shown to rapidly fuse and become damaged, and it was therefore concluded that glass transition or vitrification is an important factor in decreasing the chances of fusion upon drying (Crowe et al. 1998). Vitrification is normally achieved by combining a high concentration of CPA (high viscosity), a rapid cooling rate and a small volume (i.e. 0.1 µL) (Arav 1992; Arav and Zeron 1997). Obviously, these conditions are not feasible in the freeze-drying of many cell types because of the need to achieve a stable glass matrix without permeating CPAs, at relatively slow cooling rates and large volume.

During the dehydration process, water molecules leave the interface between the lipid bilayer, resulting in damage to the membrane and an increase in the lipid phase transition temperature. If trehalose is present in the system, it is able to replace the water molecules and keep the lipid phase transition low (Leslie et al. 1994). When water content is very low, trehalose can reach its glass transition at ambient temperatures, a state of vitrification (Buitink et al. 1998; Sun et al. 1996). The process, however, is damaging to the cellular membrane and the rehydrated cells are usually devoid of biological activity and viability. Some degree of chromosomal damage may also take place due to endogenous nucleases. This damage can be greatly reduced by the use of calcium chelating agents such as EGTA (Martins et al. 2007).

Attempts to freeze-dry spermatozoa were first reported about 6 decades ago (Polge et al. 1949). The original protocol was applied later to other species, including humans (Sherman 1954; Sherman 1957; Yushchenko 1957), but results in terms of offspring production were contradictory (Saacke and Almquist 1961; Yushchenko 1957). Experiments in fertility of freeze-dried bull spermatozoa were reported in 1976 (Larson and Graham) and since then reports on a growing number of species are accumulating in the literature. The definitive proof that freeze-dried spermatozoa retain genetic integrity was established only when microsurgical procedures for bypassing the lack of motility of freeze-dried spermatozoa were developed, and normal mice were produced from the intracytoplasmic sperm injection (ICSI) of freeze-dried sperm (Wakayama and Yanagimachi 1998). A follow-up study by the same group demonstrated the preservation of genomic integrity in freeze-dried spermatozoa (Kusakabe et al. 2001), and more recently these results have been demonstrated in other species. To date, embryonic development after ICSI with freeze-dried sperm heads has been reported in humans and hamster

(Katayose et al. 1992), cattle (Keskintepe et al. 2002; Martins et al. 2007), pigs (Kwon et al. 2004), rhesus macaque (Sanchez-Partida et al. 2008) and cats (Moisan et al. 2005), and live offspring were reported in mice (Kaneko et al. 2003; Wakayama and Yanagimachi 1998; Ward et al. 2003), rabbits (Liu et al. 2004; Yushchenko 1957), rat (Hirabayashi et al. 2005; Hochi et al. 2008) and fish (Poleo et al. 2005). Although storage at room temperature would have been the ideal solution, it seems that the storage temperature is affecting the DNA integrity. Studies on both mouse and rat spermatozoa showed that at room temperature there is progressive damage to the cells whereas at +4°C, and even more so at -196°C, these damages are reduced (Hochi et al. 2008; Kaneko and Nakagata 2005). Based on extrapolation of Arrhenius plots, there will be on going deterioration of sperm stored at +4°C but there would be no decline in the fertilizing ability (by ICSI) and the rate of development to blastocyst stage of freeze-dried spermatozoa stored at -80°C (Kawase et al. 2005). Convective drying was tested on rhesus macaque spermatogonial stem cells. After loading them with 50 mM trehalose, as many as 80% of the cells maintained viability at water content as low as 0.5g H_2O per g dry weight (Meyers 2006). The possibility to store male gametes and spermatogonial stem cells in a dry state represents a major breakthrough for storing and shipping male gametes of laboratory, farm and wild animal species as well as humans. These can be used through ICSI, which has become a routine procedure in assisted reproduction.

Unlike the seeds of food plants mentioned earlier (Ruttimann 2006), which are the equivalent of embryos in the animal kingdom, creating embryos from freeze-dried spermatozoa will require conspecific good quality oocytes and the use of ICSI to fertilize them with the rehydrated cells. An alternative to freeze-drying of gametes, which often prove very difficult to obtain, specifically in wildlife species, is freeze drying of somatic cells, to be later used for somatic cell nuclear transfer (SCNT, Wilmut et al. 1997). Somatic cell nuclear transfer, also known as cloning, has indeed an obvious potential for the multiplication of rare genotypes (Corley-Smith and Brandhorst 1999; Loi et al. 2008a; Loi et al. 2008b), but its wide application is prevented by the currently low efficiency in terms of offspring outcome. To date, successful cloning was reported in sheep (Campbell et al. 1996; Loi et al. 2008a; Loi et al. 2008b; Wilmut et al. 1997), cows (Cibelli et al. 1998), mice (Wakayama and Yanagimachi 1998), goats (Baguisi et al. 1999), pigs (Polejaeva et al. 2000), cats (Shin et al. 2002), dogs (Jang et al. 2007), rabbits (Chesne et al. 2002), ferrets (Li et al. 2006), mules (Woods et al. 2003), horses (Galli et al. 2003), gaurs (*Bos gaurus*) (Lanza et al. 2000), buffalos (*Bubalus bubalis*) (Lu et al. 2005; Shi et al. 2007), mouflons (*Ovis orientalis musimon*) (Loi et al. 2001), African wild cats (*Felis silvestris lybica*) (Gómez et al. 2003), wolves (*Canis lupus*) (Kim et al. 2007), mountain bongo antelopes (*Tragelaphus euryceros isaaci*) (Lee et al. 2003), elands (*Taurotragus oryx*) (Nel-Themaat et al. 2008), and Canada lynxes (*Lynx Canadensis*) (Gómez et al. 2009) and attempt at injecting cell nuclei from 15,000-year-old woolly mammoth into mouse oocytes was also reported recently (Kato et al. 2009).

An obvious advantage to SCNT over the use of freeze-dried spermatozoa is the possibility to generate embryos using enucleated oocytes from closely related species. As the reproductive biology of the majority of the endangered species is practically unknown, and while waiting for SCNT and other related technologies to mature and improve their efficiency, storage of somatic cells for future use is certainly a wise step to be undertaken. However, the establishment of biobanks in the form of cryopreserved cell lines encounters several problems, represented by high maintenance and liquid nitrogen costs. Recently, our group has demonstrated that somatic cells, rendered unviable by heat treatment, retained the potential to generate a normal lamb after nuclear transfer (Loi et al. 2002). We have later demonstrated that the use of sheep freeze-dried somatic cells for SCNT is a viable technique that can direct embryonic development (Loi et al. 2008a; Loi et al. 2008b). In these reports, utilizing the directional freezing technology, freeze-dried granulosa cells, kept at room temperature for 3 years, were used to direct embryonic development following nuclear transfer into *in vitro* matured enucleated oocytes. We showed in these studies that enucleated oocytes injected with freeze-dried granulosa cells initiated cleavage at the same rate as in control embryos generated using fresh granulosa cells. Microsatellite DNA analysis of the cloned blastocysts matched perfectly with the lyophilized donor cells. We have demonstrated for the first time that lyophilized cells maintain the development potential when injected into enucleated oocytes. It is also worth noting that the cells used in this study were maintained in a dehydrated state at room temperature for three years, whereas lyophilised spermatozoa were stored for only 4 months before being used for ICSI. We believe that these findings constitute a major contribution for the long-term storage of somatic cells from animals threatened by extinction. Following this lead, the ability of freeze-dried somatic cells to direct embryonic development was demonstrated in other mammals such as the pig (Das et al. 2010). In this respect, the "Dried Noah's Ark" project was initiated by us with the aim to preserve somatic cells from endangered species in the dried form. So far we have collected blood samples and freeze-dried the nucleated cells in them from 2 different antelope species, two Somali wild ass (female and male, critical endangered) (Figure 14), desert lions (critically endangered in Israel), Barbary sheep and spider monkey. The nucleated white blood cells are extracted from these blood samples and freeze-dried (Figure 15). When considering SCNT for wildlife species preservation, several important issues should be taken into consideration (Pukazhenthi and Wildt 2004 and references therein). These include: 1) availability of good quality oocytes and the ability to access them, 2) mitochondrial inheritance, specifically when oocytes from other species are used, 3) shortening of telomere following SCNT (Shiels et al. 1999), and 4) elevated prevalence of developmental abnormalities and high mortality rate following SCNT (e.g. Lanza et al. 2000). As with cryopreservation of other cells and tissues, storage space, costs and carbon footprint are major issues when liquid nitrogen is involved. The freeze-drying option can greatly reduce many of these costs as cells can be kept even at room temperature.

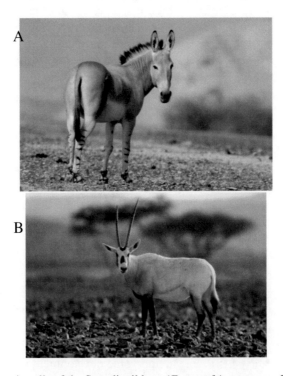

Figure 14. Somatic cells of the Somali wild ass (*Equus africanus somaliensis*) (A) and the Arabian Oryx (*Oryx leucoryx*) (B) are among the first to be stored in a freeze-dried form as part of the 'Dried Noah's Ark' project. Photos: Eyal Bartove ©.

OUR PROTOCOL FOR FREEZE-DRYING
WHITE BLOOD CELLS

Peripheral blood lymphocytes are isolate from any mammalian species through a Ficoll-Paque density gradient. The purity of the extracted cells sample can be assessed by an automatic cell counter (e.g. Pentra 60, ABX, France). The cells are frozen in a freezing solution that contains 50% fetal calf serum (FCS) and 0.1 M trehalose in Hepes-Talp buffer. Two mL samples are frozen using the multi-thermal gradient (MTG) freezing apparatus (IMT Ltd, Ness Ziona, Israel) at a cooling rate of 5.1°C/min. Cell concentration range between 1 and 10 million cells/mL. After freezing, the samples are store under liquid nitrogen until they are inserted into the lyophilizer (Freezone Plus 6, Labconco, Kansas City, MO, USA). Samples are lyophilized for 72 hours, after which, each ampoule is flame sealed, placed into a cardboard box and stored at room temperature (23 to 25°C) until use.

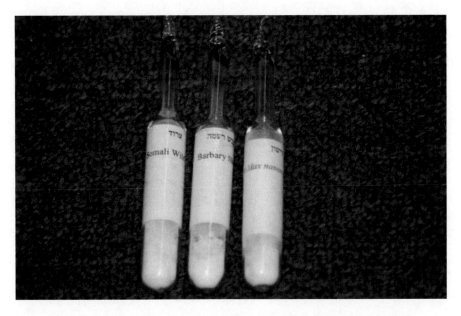

Figure 15. Ampoules with freeze-dried lymphocytes extracted from blood samples obtained from a Somali wild ass (*Equus africanus somaliensis*), a Barbary sheep (*Ammotragus lervia*) and an addax (*Addax nasomaculatus*).

REHYDRATION

Immediately before use for nuclear transfer, the ampoules are opened and 1 or 2 mL of milliQ water is added. After rehydration, cells are washed twice with medium 199 plus antibiotics and BSA before use for nuclear transfer.

FREEZE-DRYING – CONCLUSION

While freezing of sperm, oocytes and embryos, not to mention tissue slices or whole organs, is very complex and expensive technique, freeze-drying of gametes, and more so of somatic cells can raise a simple alternative technique. Both freeze-dried spermatozoa and somatic cells were shown to direct embryonic development however the utilization of both in a widespread manner is hindered by the techniques needed for their utilization. Freeze-dried spermatozoa are devoid of motility and until this hurdle has been overcome, they can only be used through ICSI, a technique that has been developed for only a handful of species. The use of freeze-dried somatic cells can only be done through SCNT and the current efficiency in the few species in which it has been attempted is very low. Freeze drying of oocytes is yet to be demonstrated. These issues, however, should not stop us from accumulating spermatozoa and somatic cells from as wide a variety of endangered species while waiting for the needed advances that will allow their used wherever and whenever needed.

REFERENCES

Abe Y, Suwa Y, Lee DS, Kim SK, Suzuki H. 2007. Vitrification of canine oocytes and embryos, and pregnancy after non-surgical tranfer of vitrified embryos. *Biology of Reproduction* 77(1 supplement):134 (abstract).

Adams GP, Ratto MH, Huanca W, Singh J. 2005. Ovulation-inducing factor in the seminal plasma of alpacas and llamas. *Biology of Reproduction* 73(3):452-457.

Allen WR. 2010. Sex, science and satisfaction: A heady brew. *Animal Reproduction Science* 121(1-2):262-278.

Aller JF, Rebuffi GE, Cancino AK, Alberio RH. 2002. Successful transfer of vitrified llama (*Lama glama*) embryos. *Animal Reproduction Science* 73(1-2):121-127.

Allison SD, Randolph TW, Manning MC, Middleton K, Davis A, Carpenter JF. 1998. Effects of drying methods and additives on structure and function of actin: Mechanisms of dehydration-induced damage and its inhibition. *Archives of Biochemistry and Biophysics* 358(1):171-181.

Almodin CG, Minguetti-Camara VC, Meister H, Ferreira JOHR, Franco RL, Cavalcante AA, Radaelli MRM, Bahls AS, Moron AF, Murta CGV. 2004. Recovery of fertility after grafting of cryopreserved germinative tissue in female rabbits following radiotherapy. *Human Reproduction* 19(6):1287-1293.

Almodin CG, Minguetti-Camara VC, Paixao CL, Pereira PC. 2010. Embryo development and gestation using fresh and vitrified oocytes. *Human Reproduction* 25(5):1192-1198.

Andrabi SMH, Maxwell WMC. 2007. A review on reproductive biotechnologies for conservation of endangered mammalian species. *Animal Reproduction Science* 99(3-4):223-243.

Arav A. 1992. Vitrification of oocytes and embryos. In: A. L, Gandolfi F, editors. *New Trends in Embryo Transfer.* Cambridge, UK: Portland Press. p 255-264.

Arav A; Arav, Amir, assignee. 1999. *Device and methods for multigradient directional cooling and warming of biological samples.* US Patent patent 5,873,254.

Arav A. 2001. Transillumination increases oocyte recovery from ovaries collected at slaughter. A new technique report. *Theriogenology* 55(7):1561-1565.

Arav A. 2003. Large tissue freezing. *Journal of Assisted Reproduction and Genetics* 20(9):351.

Arav A, Elami A. 2005. Cryopreservation of intact ovaries--size is a variable? Reply of the Authors. *Fertility and Sterility* 83(5):1588.

Arav A, Gavish Z, Elami A, Natan Y, Revel A, Silber S, Gosden RG, Patrizio P. 2010. Ovarian function 6 yers after cryopreservation and transplantation of whole sheep ovaries. *Reproductive Biomedicine Online* 20(1):48.

Arav A, Natan Y. 2009. Directional freezing: a solution to the methodological challenges to preserve large organs. *Seminars in Reproductive Medicine* 27(6):438-442.

Arav A, Pearl M, Zeron Y. 2000a. Does lipid profile explains chilling sensitivity and membrane lipid phase transition of spermatozoa and oocytes? *CryoLetters* 21:179-186.

Arav A, Shehu D, Mattioli M. 1993. Osmotic and cytotoxic study of vitrification of immature bovine oocytes. *Journal of Reproduction and Fertility* 99(2):353-358.

Arav A, Yavin S, Zeron Y, Natan D, Dekel I, Gacitua H. 2002a. New trends in gamete's cryopreservation. *Molecular and Cellular Endocrinology* 187(1-2):77-81.

Arav A, Zeron Y. 1997. Vitrification of bovine oocytes using modified minimum drop size technique (MDS) is effected by the composition and the concentration of the vitrification solution and by the cooling conditions. *Theriogenology* 47(1):341 (abstract).

Arav A, Zeron Y, Leslie SB, Behboodi E, Anderson GB, Crowe JH. 1996. Phase transition temperature and chilling sensitivity of bovine oocytes. *Cryobiology* 33(6):589-599.

Arav A, Zeron Y, Ocheretny A. 2000b. A new device and method for vitrification increases the cooling rate and allows successful cryopreservation of bovine oocytes. *Theriogenology* 53(1):248 (abstract).

Arav A, Zeron Y, Shturman H, Gacitua H. 2002b. Successful pregnancies in cows following double freezing of a large volume of semen. *Reproduction, Nutrition, Development* 42(6):583-586.

Armstrong DL, Looney CR, Lindsey BR, Gonseth CL, Johnson DL, Williams KR, Simmons LG, Loskutoff NM. 1995. Transvaginal egg retrieval and *in-vitro* embryo production in gaur (*Bos gaurus*) with establishment of interspecies pregnancy. *Theriogenology* 43(1):162 (abstract).

Asada M, Horii M, Mogoe T, Fukui Y, Ishikawa H, Ohsumi S. 2000. *In vitro* maturation and ultrastructural observation of cryopreserved minke whale (*Balaenoptera acutorostrata*) follicular oocytes. *Biology of Reproduction* 62(2):253-259.

Asada M, Tetsuka M, Ishikawa H, Ohsumi S, Fukui Y. 2001a. Improvement on *in vitro* maturation, fertilization and development of minke whale (*Balaenoptera acutorostrata*) oocytes. *Theriogenology* 56(4):521-533.

Asada M, Wei H, Nagayama R, Tetsuka M, Ishikawa H, Ohsumi S, Fukui Y. 2001b. An attempt at intracytoplasmic sperm injection of frozen-thawed minke whale (*Balaenoptera bonaerensis*) oocytes. *Zygote* 9(4):299-307.

Asher GW, Berg DK, Evans G. 2000. Storage of semen and artificial insemination in deer. *Animal Reproduction Science* 62(1-3):195-211.

Avarbock MR, Brinster CJ, Brinster RL. 1996. Reconstitution of spermatogenesis from frozen spermatogonial stem cells. *Nature Medicine* 2(6):693-696.

Bagis H, Akkoc T, Tass A, Aktoprakligil D. 2008. Cryogenic effect of antifreeze protein on transgenic mouse ovaries and the production of live offspring by orthotopic transplantation of cryopreserved mouse ovaries. *Molecular Reproduction and Development* 75(4):608-613.

Baguisi A, Behboodi E, Melican DT, Pollock JS, Destrempes MM, Cammuso C, Williams JL, Nims SD, Porter CA, Midura P and others. 1999. Production of goats by somatic cell nuclear transfer. *Nature Biotechnology* 17(5):456-461.

Baird DT, Webb R, Campbell BK, Harkness LM, Gosden RG. 1999. Long-term ovarian function in sheep after ovariectomy and transplantation of autografts stored at -196 C. *Endocrinology* 140(1):462-471.

Ball BA, Medina V, Gravance CG, Baumber J. 2001. Effect of antioxidants on preservation of motility, viability and acrosomal integrity of equine spermatozoa during storage at 5°C. *Theriogenology* 56(4):577-589.

Ballou JD. 1992. Potential contribution of cryopreserved germ plasm to the preservation of genetic diversity and conservation of endangered species in captivity. *Cryobiology* 29(1):19-25.

Balmaceda JP, Heitman TO, Garcia MR, Pauerstein CJ, Pool TB. 1986. Embryo cryopreservation in cynomolgus monkeys. *Fertility and Sterility* 45(3):403-406.

Balmaceda JP, Pool TB, Arana JB, Heitman TS, Asch RH. 1984. Successful *in vitro* fertilization and embryo transfer in cynomolgus monkeys. *Fertility and Sterility* 42(5):791-795.

Bandularatne E, Bongso A. 2002. Evaluation of human sperm function after repeated freezing and thawing. *Journal of Andrology* 23(2):242-249.

Bank H, Maurer RR. 1974. Survival of frozen rabbit embryos. *Experimental Cell Research* 89(1):188-196.

Barfield JP, McCue PM, Squires EL, Seidel GE, Jr. 2009. Effect of dehydration prior to cryopreservation of large equine embryos. *Cryobiology* 59(1):36-41.

Barone MA, Roelke ME, Howard J, Brown JL, Anderson AE, Wildt DE. 1994. Reproductive characteristics of male Florida panthers: Comparative studies from Florida, Texas, Colorado, Latin America, and North American zoos. *Journal of Mammalogy* 75(1):150-162.

Bart AN, Kyaw HA. 2003. Survival of zebrafish, *Brachydanio rerio* (Hamilton-Buchanan), embryo after immersion in methanol and exposure to ultrasound with implications to cryopreservation. *Aquaculture Research* 34(8):609-615.

Bartels P, Kotze A. 2006. Wildlife biomaterial banking in Africa for now and the future. *Journal of Environmental Monitoring* 8(8):779-781.

Baumber J, Ball BA, Gravance CG, Medina V, Davies-Morel MC. 2000. The effect of reactive oxygen species on equine sperm motility, viability, acrosomal integrity, mitochondrial membrane potential, and membrane lipid peroxidation. *Journal of Andrology* 21(6):895-902.

Bedaiwy MA, Hussein MR, Biscotti C, Falcone T. 2006. Cryopreservation of intact human ovary with its vascular pedicle. *Human Reproduction* 21(12):3258-3269.

Beebe LFS, Cameron RDA, Blackshaw AW, Keates HL. 2005. Changes to porcine blastocyst vitrification methods and improved litter size after transfer. *Theriogenology* 64(4):879-890.

Behr B, Rath D, Hildebrandt TB, Goeritz F, Blottner S, Portas TJ, Bryant BR, Sieg B, Knieriem A, de Graaf SP and others. 2009a. Germany/Australia index of sperm sex sortability in elephants and rhinoceros. *Reproduction in Domestic Animals* 44(2):273-277.

Behr B, Rath D, Mueller P, Hildebrandt TB, Goeritz F, Braun BC, Leahy T, de Graaf SP, Maxwell WMC, Hermes R. 2009b. Feasibility of sex-sorting

References 93

sperm from the white and the black rhinoceros (*Ceratotherium simum, Diceros bicornis*). *Theriogenology* 72(3):353-364.

Berg DK, Asher GW. 2003. New developments reproductive technologies in deer. *Theriogenology* 59(1):189-205.

Berlinguer F, Gonzalez R, Succu S, del Olmo A, Garde JJ, Espeso G, Gomendio M, Ledda S, Roldan ER. 2008. *In vitro* oocyte maturation, fertilization and culture after ovum pick-up in an endangered gazelle (*Gazella dama mhorr*). *Theriogenology* 69(3):349-359.

Bielanski A, Bergeron H, Lau PC, Devenish J. 2003. Microbial contamination of embryos and semen during long term banking in liquid nitrogen. *Cryobiology* 46(2):146-152.

Bielanski A, Vajta G. 2009. Risk of contamination of germplasm during cryopreservation and cryobanking in IVF units. *Human Reproduction* 24(10):2457-2467.

Bilton RJ, Moore NW. 1976. *In vitro* culture, storage and transfer of goat embryos. *Australian Journal of Biological Sciences* 29(1-2):125-129.

Blesbois E, Lessire M, Grasseau I, Hallouis J, Hermier D. 1997. Effect of dietary fat on the fatty acid composition and fertilizing ability of fowl semen. *Biology of Reproduction* 56(5):1216-1220.

Blondin P, Coenen K, Guilbault LA, Sirard MA. 1996. Superovulation can reduce the developmental competence of bovine embryos. *Theriogenology* 46(7):1191-1203.

Borini A, Sereni E, Bonu MA, Flamigni C. 2000. Freezing a few testicular spermatozoa retrieved by TESA. *Molecular and Cellular Endocrinology* 169(1-2):27-32.

Bosch P, Hernandez-Fonseca HJ, Miller DM, Wininger JD, Massey JB, Lamb SV, Brackett BG. 2004. Development of antral follicles in cryopreserved cat ovarian tissue transplanted to immunodeficient mice. *Theriogenology* 61(2-3):581-594.

Bouamama N, Briot P, Testart J. 2003. [Comparison of two methods of cryoconservation of sperm when in very small numbers]. *Gynécologie Obstétrique & Fertilité* 31(2):132-135 (French).

Boutelle S, Lenahan K, Krisher R, Bauman KL, Asa CS, Silber S. 2010. Vitrification of oocytes from endangered Mexican gray wolves (*Canis lupus baileyi*). *Theriogenology* 75(4):647-654.

Bowen RA. 1977. Fertilization *in vitro* of feline ova by spermatozoa from the ductus deferens. *Biology of Reproduction* 17(1):144-147.

Bravo PW, Skidmore JA, Zhao XX. 2000. Reproductive aspects and storage of semen in Camelidae. *Animal Reproduction Science* 62(1-3):173-193.

Breed WG. 1994. How does sperm meet egg?--in a marsupial. *Reproduction, Fertility and Development* 6(4):485-506.

Breed WG, Taggart DA, Bradtke V, Leigh CM, Gameau L, Carroll J. 1994. Effect of cryopreservation on development and ultrastructure of preimplantation embryos from the dasyurid marsupial *Sminthopsis crassicaudata. Journal of Reproduction and Fertility* 100(2):429-438.

Brinsko SP, Varner DD, Love CC, Blanchard TL, Day BC, Wilson ME. *Effect of feeding a DHA-enriched nutriceutical on motion characteristics of cooled and frozen stallion semen;* 2003; New Orleans, Louisiana. www.ivis.org Document No. P0653.1103.

Brinster RL, Zimmermann JW. 1994. Spermatogenesis following male germ-cell transplantation. *Proceedings of the National Academy of Sciences of the United States of America* 91(24):11298-11302.

Bromer JG, Patrizio P. 2009. Fertility preservation: the rationale for cryopreservation of the whole ovary. *Seminars in Reproductive Medicine* 27(6):465-471.

Brown JL, Wildt DE, Wielebnowski N, Goodrowe KL, Graham LH, Wells S, Howard JG. 1996. Reproductive activity in captive female cheetahs (*Acinonyx jubatus*) assessed by faecal steroids. *Journal of Reproduction and Fertility* 106(2):337-346.

Brown JL, Wildt DE. 1997. Assessing reproductive status in wild felids by noninvasive faecal steroid monitoring. *International Zoo Yearbook* 35(1):173-191.

Buitink J, Claessens MM, Hemminga MA, Hoekstra FA. 1998. Influence of water content and temperature on molecular mobility and intracellular glasses in seeds and pollen. *Plant Physiology* 118(2):531-41.

Cabrita E, Robles V, Wallace JC, Sarasquete MC, Herr·ez MP. 2006. Preliminary studies on the cryopreservation of gilthead seabream (*Sparus aurata*) embryos. *Aquaculture* 251(2-4):245-255.

Cai XY, Chen GA, Lian Y, Zheng XY, Peng HM. 2005. Cryoloop vitrification of rabbit oocytes. *Human Reproduction* 20(7):1969-1974.

Campbell KH, McWhir J, Ritchie WA, Wilmut I. 1996. Sheep cloned by nuclear transfer from a cultured cell line. *Nature* 380(6569):64-66.

Camus A, Clairaz P, Ersham A, Van Kappel AL, Savic G, Staub C. 2006. [Principe de la vitrification: cinétiques comparatives. The comparison of the process of five different vitrification devices.] *Gynécologie Obstétrique & Fertilité* 34(9):737-745. (French)

Candy CJ, Wood MJ, Whittingham DG. 1995. Ovary and ovulation: Follicular development in cryopreserved marmoset ovarian tissue after transplantation. *Human Reproduction* 10(9):2334-2338.

Candy CJ, Wood MJ, Whittingham DG. 1997. Effect of cryoprotectants on the survival of follicles in frozen mouse ovaries. *Journal of Reproduction and Fertility* 110(1):11-19.

Carroll J, Gosden RG. 1993. Transplantation of frozen-thawed mouse primordial follicles. *Human Reproduction* 8(8):1163-1167.

Carroll J, Whittingham DG, Wood MJ, Telfer E, Gosden RG. 1990. Extra-ovarian production of mature viable mouse oocytes from frozen primary follicles. *Journal of Reproduction and Fertility* 90(1):321-327.

Ceballos G, Ehrlich PR. 2002. Mammal population losses and the extinction crisis. *Science* 296(5569):904-907.

Chang HJ, Lee JR, Chae SJ, Jee BC, Suh CS, Kim SH. 2008. Comparative study of two cryopreservation methods of human spermatozoa: *vitrification versus slow freezing. Fertility and Sterility* 90(Supplement 1):S280 (abstract).

Chapin FSI, Zavaleta ES, Eviner VT, Naylor RL, Vitousek PM, Reynolds HL, Hooper DU, Lavorel S, Sala OE, Hobbie SE and others. 2000. Consequences of changing biodiversity. *Nature* 405(6783):234-242.

Chen C. 1986. Pregnancy after human oocyte cryopreservation. *Lancet* 1(8486):884-886.

Chen C, Chen S, Chang F, Wu G, Liu J, Yu C. 2005. Autologous heterotropic transplantation of intact rabbit ovary after cryopreservation. *Human Reproduction* 20(suppl. 1):i149-i150 (abstract).

Chen C-H, Chen S-G, Wu G-J, Wang J, Yu C-P, Liu J-Y. 2006a. Autologous heterotopic transplantation of intact rabbit ovary after frozen banking at -196°C. *Fertility and Sterility* 86(4, Supplement 1):1059-1066.

Chen H, Cheung MP, Chow PH, Cheung ALM, Liu W, O W-S. 2002. Protection of sperm DNA against oxidative stress *in vivo* by accessory sex gland secretions in male hamsters. *Reproduction* 124(4):491-499.

Chen H-I, Tsai C-D, Wang H-T, Hwang S-M. 2006b. Cryovial with partial membrane sealing can prevent liquid nitrogen penetration in submerged storage. *Cryobiology* 53(2):283-287.

Chen LM, Hou R, Zhang ZH, Wang JS, An XR, Chen YF, Zheng HP, Xia GL, Zhang MJ. 2007. Electroejaculation and semen characteristics of Asiatic Black bears (*Ursus thibetanus*). *Animal Reproduction Science* 101(3-4):358-364.

Chen S-U, Chien C-L, Wu M-Y, Chen T-H, Lai S-M, Lin C-W, Yang Y-S. 2006c. Novel direct cover vitrification for cryopreservation of ovarian tissues increases follicle viability and pregnancy capability in mice. *Human Reproduction* 21(11):2794-2800.

Chen S-U, Yang Y-S. 2009. Slow freezing or vitrification of oocytes: Their effects on survival and meiotic spindles, and the time schedule for clinical practice. *Taiwanese Journal of Obstetrics and Gynecology* 48(1):15-22.

Chen SL, Tian YS. 2005. Cryopreservation of flounder (*Paralichthys olivaceus*) embryos by vitrification. *Theriogenology* 63(4):1207-1219.

Chen SU, Lien YR, Cheng YY, Chen HF, Ho HN, Yang YS. 2001. Vitrification of mouse oocytes using closed pulled straws (CPS) achieves a high survival and preserves good patterns of meiotic spindles, compared with conventional straws, open pulled straws (OPS) and grids. *Human Reproduction* 16(11):2350-2356.

Chesne P, Adenot PG, Viglietta C, Baratte M, Boulanger L, Renard JP. 2002. Cloned rabbits produced by nuclear transfer from adult somatic cells. *Nature Biotechnology* 20(4):366-369.

Chian RC, Son WY, Huang JY, Cui SJ, Buckett WM, Tan SL. 2005. High survival rates and pregnancies of human oocytes following vitrification: preliminary report. *Fertility and Sterility* 84(Supplement 1):S36 (abstract).

Choi YH, Hartman DL, Bliss SB, Hayden SS, Blanchard TL, Hinrichs K. 2009. High pregnancy rates after transfer of large equine blastocysts collapsed via micromanipulation before vitrification. *Reproduction, Fertility and Development* 22(1):203 (abstract).

Cibelli JB, Stice SL, Golueke PJ, Kane JJ, Jerry J, Blackwell C, Ponce de Leon FA, Robl JM. 1998. Cloned transgenic calves produced from nonquiescent fetal fibroblasts. *Science* 280(5367):1256-1258.

Ciotti PM, Porcu E, Notarangelo L, Magrini O, Bazzocchi A, Venturoli S. 2009. Meiotic spindle recovery is faster in vitrification of human oocytes compared to slow freezing. *Fertility and Sterility* 91(6):2399-2407.

Clarke AG. 2009. The Frozen Ark Project: the role of zoos and aquariums in preserving the genetic material of threatened animals. *International Zoo Yearbook* 43(1):222-230.

Clarke GN, Liu DY, Baker HWG. 2003. Improved sperm cryopreservation using cold cryoprotectant. Reproduction, *Fertility and Development* 15(7):377-381.

Cleary M, Paris MC, Shaw J, Jenkin G, Trounson A. 2003. Effect of ovariectomy and graft position on cryopreserved common wombat

(*Vombatus ursinus*) ovarian tissue following xenografting to nude mice. *Reproduction Fertility and Development* 15(6):333-342.

Cohen J, Garrisi GJ. 1997. Micromanipulation of gametes and embryos: Cryopreservation of a single human spermatozoon within an isolated zona pellucida. *Human Reproduction Update* 3(5):453.

Cohen J, Garrisi GJ, Congedo-Ferrara TA, Kieck KA, Schimmel TW, Scott RT. 1997. Cryopreservation of single human spermatozoa. *Human Reproduction* 12(5):994-1001.

Coloma MA, Toledano-Dìaz A, LÛpez-Sebasti·n A, Santiago-Moreno J. 2010. The influence of washing Spanish ibex (*Capra pyrenaica*) sperm on the effects of cryopreservation in dependency of the photoperiod. *Theriogenology* 73(7):900-908.

Combelles CMH, Racowsky C. 2005. Assessment and optimization of oocyte quality during assisted reproductive technology treatment. *Seminars in Reproductive Medicine* 23(3):277-284.

Comhaire FH, Christophe AB, Zalata AA, Dhooge WS, Mahmoud AMA, Depuydt CE. 2000. The effects of combined conventional treatment, oral antioxidants and essential fatty acids on sperm biology in subfertile men. *Prostaglandins, Leukotrienes and Essential Fatty Acids* 63(3):159-165.

Comizzoli P, Mermillod P, Mauget R. 2000. Reproductive biotechnologies for endangered mammalian species. *Reproduction Nutrition Development* 40:493-504.

Comizzoli P, Wildt DE, Pukazhenthi BS. 2003. Overcoming poor *in vitro* nuclear maturation and developmental competence of domestic cat oocytes during the non-breeding season. *Reproduction* 126(6):809-816.

Contri A, De Amicis I, Veronesi MC, Faustini M, Robbe D, Carluccio A. 2010. Efficiency of different extenders on cooled semen collected during long and short day length seasons in Martina Franca donkey. *Animal Reproduction Science* 120(1-4):136-141.

Convention on Biological Diversity. 2002. *Strategic Plan for the Convention on Biological Diversity*. Decision VI/26 of the Conference of the Parties to the Convention on Biological Diversity: United Nations Environment Programme (UNEP). http://www.cbd.int/decision/cop/?id=7200 www.cbd.int/decisions, Accessed on 13 January, 2010.

Corley-Smith GE, Brandhorst BP. 1999. Preservation of endangered species and populations: a role for genome banking, somatic cell cloning, and androgenesis? *Molecular Reproduction and Development* 53(3):363-367.

Coticchio G, Sereni E, Serrao L, Mazzone S, Iadarola I, Borini A. 2004. What criteria for the definition of oocyte quality? *Annals of the New York*

Academy of Sciences 1034(The Uterus and Human Reproduction):132-144.

Courbiere B, Caquant L, Mazoyer C, Franck M, Lornage J, Salle B. 2009. Difficulties improving ovarian functional recovery by microvascular transplantation and whole ovary vitrification. *Fertility and Sterility* 91(6):2697-2706.

Cowley CW, Timson WJ, Sawdye JA. 1961. Ultra rapid cooling techniques in the freezing of biological materials. *Biodynamica* 8(170):317-329.

Crabbe E, Verheyen G, Tournaye H, Van Steirteghem A. 1999. Freezing of testicular tissue as a minced suspension preserves sperm quality better than whole-biopsy freezing when glycerol is used as cryoprotectant. *International Journal of Andrology* 22(1):43-48.

Cranfield MR, Berger NG, Kempske S, Bavister BD, Boatman DE, Ialeggio DM. 1992. Macaque monkey birth following transfer of *in vitro* fertilized, frozen-thawed embryos to a surrogate mother. *Theriogenology* 37(1):197 (abstract).

Criado E, Albani E, Novara PV, Smeraldi A, Cesana A, Parini V, Levi-Setti PE. 2010. Human oocyte ultravitrification with a low concentration of cryoprotectants by ultrafast cooling: a new protocol. *Fertility and Sterility* 95(3):1101-1103.

Crichton EG, Armstrong DL, Vajta G, Pope CE, Loskutoff NM. 2000. Developmental competence *in vitro* of embryos produced from Siberian tigers (*Panthera tigris altaica*) cryopreserved by controlled rate freezing versus vitrification. *Theriogenology* 53(1):328 (abstract).

Crichton EG, Bedows E, Miller-Lindholm AK, Baldwin DM, Armstrong DL, Graham LH, Ford JJ, Gjorret JO, Hyttel P, Pope CE and others. 2003. Efficacy of porcine gonadotropins for repeated stimulation of ovarian activity for oocyte retrieval and *in vitro* embryo production and cryopreservation in Siberian tigers (*Panthera tigris altaica*). *Biology of Reproduction* 68(1):105-113.

Crosier AE, Pukazhenthi BS, Henghali JN, Howard J, Dickman AJ, Marker L, Wildt DE. 2006. Cryopreservation of spermatozoa from wild-born Namibian cheetahs (*Acinonyx jubatus*) and influence of glycerol on cryosurvival. *Cryobiology* 52(2):169-181.

Crowe JH, Carpenter JF, Crowe LM. 1998. The role of vitrification in anhydrobiosis. *Annual Review of Physiology* 60(1):73-103.

Crowe JH, Crowe LM, Chapman D. 1984. Preservation of membranes in anhydrobiotic organisms: The role of trehalose. *Science* 223(4637):701-703.

Crowe JH, Oliver AE, Hoekstra FA, Crowe LM. 1997. Stabilization of dry membranes by mixtures of hydroxyethyl starch and glucose: the role of vitrification. *Cryobiology* 35(1):20-30.

Crowe LM, Crowe JH. 1991. Solution effects on the thermotropic phase transition of unilamellar liposomes. *Biochimica et Biophysica Acta (BBA) - Biomembranes* 1064(2):267-274.

Cuello C, Gil MA, Parrilla I, Tornel J, Vázquez JM, Roca J, Berthelot F, Martinat-Botté F, Martínez EA. 2004. Vitrification of porcine embryos at various developmental stages using different ultra-rapid cooling procedures. *Theriogenology* 62(1-2):353-361.

Culver JN. 2001. Evaluation of tom fertility as affected by dietary fatty acid composition [Master of Science]. Blacksburg: Virginia Polytechnic Institute and State University. 60 p.

Curnow EC, Kuleshova LL, Shaw JM, Hayes ES. 2002. Comparison of slow- and rapid-cooling protocols for early-cleavage-stage *Macaca fascicularis* embryos. *American Journal of Primatology* 58(4):169-174.

Czarny NA, Harris MS, Rodger JC. 2009. Dissociation and preservation of preantral follicles and immature oocytes from female dasyurid marsupials. *Reproduction, Fertility and Development* 21(5):640-648.

Das ZC, Gupta MK, Uhm SJ, Lee HT. 2010. Lyophilized somatic cells direct embryonic development after whole cell intracytoplasmic injection into pig oocytes. *Cryobiology* 61(2):220-224.

de Andrade AFC, Zaffalon FG, Celeghini ECC, Nascimento J, Tarragó OFB, Martins S, Alonso MA, Arruda RP. 2010. Addition of seminal plasma to post-thawing equine semen: What is the effect on sperm cell viability? *Reproduction in Domestic Animals* In Press.

de Graaf SP, Evans G, Maxwell WMC, Cran DG, O'Brien JK. 2007. Birth of offspring of pre-determined sex after artificial insemination of frozen-thawed, sex-sorted and re-frozen-thawed ram spermatozoa. *Theriogenology* 67(2):391-398.

Desai N, Culler C, Goldfarb J. 2004a. Cryopreservation of single sperm from epididymal and testicular samples on cryoloops: Preliminary case report. *Fertility and Sterility* 82(Supplement 2):S264-S265 (abstract).

Desai NN, Blackmon H, Goldfarb J. 2004b. Single sperm cryopreservation on cryoloops: an alternative to hamster zona for freezing individual spermatozoa. *Reproduction Biomedicine Online* 9(1):47-53.

Devroey P, Liu J, Nagy Z, Goossens A, Tournaye H, Camus M, van Steirteghem A, Silber S. 1995. Pregnancies after testicular sperm extraction and intracytoplasmic sperm injection in non-obstructive azoospermia. *Human Reproduction* 10(6):1457-1460.

Ding FH, Xiao ZZ, Li J. 2007. Preliminary studies on the vitrification of red sea bream (*Pagrus major*) embryos. *Theriogenology* 68(5):702-708.

Dinnyes A, Dai Y, Jiang S, Yang X. 2000. High developmental rates of vitrified bovine oocytes following parthenogenetic activation, *in vitro* fertilization, and somatic cell nuclear transfer. *Biology of Reproduction* 63(2):513-518.

Dixon TE, Hunter JW, Beatson NS. 1991. Pregnancies following the export of frozen red deer embryos from New Zealand to Australia. *Theriogenology* 35(1):193 (Abstract).

Dobrinsky JR, Pursel VG, Long CR, Johnson LA. 2000. Birth of piglets after transfer of embryos cryopreserved by cytoskeletal stabilization and vitrification. *Biology of Reproduction* 62(3):564-570.

Donnez J, Dolmans MM, Demylle D, Jadoul P, Pirard C, Squifflet J, Martinez-Madrid B, van Langendonckt A. 2004. Livebirth after orthotopic transplantation of cryopreserved ovarian tissue. *Lancet* 364(9443):1405-1410.

Donoghue AM, Johnston LA, Seal US, Armstrong DL, Tilson RL, Wolf P, Petrini K, Simmons LG, Gross T, Wildt DE. 1990. *In vitro* fertilization and embryo development *in vitro* and *in vivo* in the tiger (*Panthera tigris*). *Biology of Reproduction* 43(5):733-744.

Dresser BL, Gelwicks EJ, Wachs KB, Keller GL. 1988. First successful transfer of cryopreserved feline (*Felis catus*) embryos resulting in live offspring. *Journal of Experimental Zoology* 246(2):180-186.

Dresser BL, Kramer L, Dalhausen RD, Pope CE, Baker RD. Cryopreservation followed by successful embryo transfer of African eland antelope; 1984 June 10-14, 1984; University of Illinois at Urbana-Champaign, Illinois, USA. p 191-193.

Dresser BL, Pope CE, Kramer L, Kuehn G, Dahlhausen RD, Maruska EJ, Reece B, Thomas WD. 1985. Birth of bongo antelope (*Tragelaphus euryceros*) to eland antelope (*Tragelaphus oryx*) and cryopreservation of bongo embryos. *Theriogenology* 23(1):190 (abstract).

Drobnis EZ, Crowe LM, Berger T, Anchordoguy TJ, Overstreet JW, Crowe JH. 1993. Cold shock damage is due to lipid phase transitions in cell membranes: A demonstration using sperm as a model. *Journal of Experimental Zoology* 265:432-437.

Du Y, Pribenszky CS, Molnar M, Zhang X, Yang H, Kuwayama M, Pedersen AM, Villemoes K, Bolund L, Vajta G. 2008. High hydrostatic pressure: a new way to improve *in vitro* developmental competence of porcine matured oocytes after vitrification. *Reproduction* 135(1):13-17.

Duraiappah AK, Naeem S, Agardy T, Ash NJ, Cooper HD, Díaz S, Faith DP, Mace G, McNeely JA, Mooney HA and others. 2005. *Ecosystems and Human Well-being: Biodiversity Synthesis*. Washington, DC: World Resources Institute. VI, 86 p.

Durrant BS. 1983. Reproductive studies of the Oryx. *Zoo Biology* 2(3):191-197.

Edgar DH, Gook DA. 2007. How should the clinical efficiency of oocyte cryopreservation be measured? *Reproduction Biomedicine Online* 14(4):430-435.

Ehmcke J, Schlatt S. 2008. Animal models for fertility preservation in the male. *Reproduction* 136(6):717-723.

Ehrlich PR, Wilson EO. 1991. Biodiversity studies: science and policy. *Science* 253(5021):758-762.

Eppig JJ, O'Brien MJ. 1996. Development *in vitro* of mouse oocytes from primordial follicles. *Biology of Reproduction* 54(1):197-207.

Erwin DH. 2001. Lessons from the past: biotic recoveries from mass extinctions. *Proceedings of the National Academy of Science of the Uunited States of America* 98(10):5399-5403.

Escriba M-J, Grau N, Escrich L, Pellicer A. 2010. Vitrification of isolated human blastomeres. *Fertility and Sterility* 93(2):669-671.

Fabbri R, Porcu E, Marsella T, Primavera MR, Rocchetta G, Ciotti PM, Magrini O, Seracchioli R, Venturoli S, Flamigni C. 2000. Technical aspects of oocyte cryopreservation. *Molecular and Cellular Endocrinology* 169(1-2):39-42.

Fahy GM, Saur J, Williams RJ. 1990. Physical problems with the vitrification of large biological systems. *Cryobiology* 27(5):492-510.

Farstad W. 2000a. Assisted reproductive technology in canid species. *Theriogenology* 53(1):175-186.

Farstad W. 2000b. Current state in biotechnology in canine and feline reproduction. *Animal Reproduction Science* 60-61:375-387.

Feng L-X, Chen Y, Dettin L, Pera RAR, Herr JC, Goldberg E, Dym M. 2002. Generation and *in vitro* differentiation of a spermatogonial cell line. *Science* 297(5580):392-395.

Filliers M, Rijsselaere T, Bossaert P, Zambelli D, Anastasi P, Hoogewijs M, Van Soom A. 2010. *In vitro* evaluation of fresh sperm quality in tomcats: A comparison of two collection techniques. *Theriogenology* 74(1):31-39.

Fiser PS, Fairfull RW. 1986. The effects of rapid cooling (cold shock) of ram semen, photoperiod, and egg yolk in diluents on the survival of spermatozoa before and after freezing. *Cryobiology* 23(6):518-524.

Freeman EW, Abbondanza FN, Meyer JM, Schulte BA, Brown JL. 2010. A simplified method for monitoring progestagens in African elephants under field conditions. *Methods in Ecology and Evolution* 1(1):86-91.

Fujihira T, Kobayashi M, Hochi S, Hirabayashi M, Ishikawa H, Ohsumi S, Fukui Y. 2006. Developmental capacity of Antarctic minke whale (*Balaenoptera bonaerensis*) vitrified oocytes following *in vitro* maturation, and parthenogenetic activation or intracytoplasmic sperm injection. *Zygote* 14(2):89-95.

Fusi F, Calzi F, Rabellotti E, Papaleo E, Gonfiantini C, Bonzi V, De Santis L, Ferrari A. 2001. Fertilizing capability of frozen–thawed spermatozoa, recovered from microsurgical epididymal sperm aspiration and cryopreserved in oocyte-free human zona pellucida. *Human Reproduction* 16(suppl 1):117-118 (abstract).

Gacitua H, Arav A. 2005. Successful pregnancies with directional freezing of large volume buck semen. *Theriogenology* 63(3):931-938.

Gacitua H, Zeron Y, Arav A. 2002. Dietary supplementation with polyunsaturated fatty acid modifies sperm quality and fatty acid composition in the bull. *Theriogenology* 57(1):376 (abstract).

Galli C, Lagutina I, Crotti G, Colleoni S, Turini P, Ponderato N, Duchi R, Lazzari G. 2003. A cloned horse born to its dam twin. *Nature* 424:635.

Gandolfi F, Paffoni A, Papasso Brambilla E, Bonetti S, Brevini TA, Ragni G. 2006. Efficiency of equilibrium cooling and vitrification procedures for the cryopreservation of ovarian tissue: comparative analysis between human and animal models. *Fertility and Sterility* 85(Suppl. 1):1150-1156.

Garner DL. 2006. Flow cytometric sexing of mammalian sperm. *Theriogenology* 65(5):943-957.

Gastal MO, Henry M, Beker AR, Gastal EL, Goncalves A. 1996. Sexual behavior of donkey jacks: Influence of ejaculatory frequency and season. *Theriogenology* 46(4):593-603.

Ghetler Y, Yavin S, Shalgi R, Arav A. 2005. The effect of chilling on membrane lipid phase transition in human oocytes and zygotes. *Human Reproduction* 20(12):3385-3389.

Gianaroli L, Magli MC, Selman HA, Colpi G, Belgrano E, Trombetta C, Vitali G, Ferraretti AP. 1999. Diagnostic testicular biopsy and cryopreservation of testicular tissue as an alternative to repeated surgical openings in the treatment of azoospermic men. *Human Reproduction* 14(4):1034-1038.

Gibson CD, Graham EF. 1969. The relationship between fertility and post-freeze motility of bull spermatozoa (by pellet freezing) without glycerol. *Journal of Reproduction and Fertility* 20(1):155-157.

Gil-Salom M, Romero J, Rubio C, Ruiz A, Remohì J, Pellicer A. 2000. Intracytoplasmic sperm injection with cryopreserved testicular spermatozoa. *Molecular and Cellular Endocrinology* 169(1-2):15-19.

Gillan L, Maxwell WMC, Evans G. 2004. Preservation and evaluation of semen for artificial insemination. *Reproduction, Fertility and Development* 16:447-454.

Gomendio M, Cassinello J, Roldan ERS. 2000. A comparative study of ejaculate traits in three endangered ungulates with different levels of inbreeding: fluctuating asymmetry as an indicator of reproductive and genetic stress. *Proceedings of the Royal Society of London. Series B: Biological Sciences* 267(1446):875-882.

Gómez MC, Jenkins JA, Giraldo A, Harris RF, King A, Dresser BL, Pope CE. 2003. Nuclear transfer of synchronized African wild cat somatic cells into enucleated domestic cat oocytes. *Biology of Reproduction* 69(3):1032-1041.

Gómez MC, Lyons JI, Pope CE, Biancardi M, Dumas C, Galiguis J, Wang G, Dresser BL. 2009. Effects of phylogenic genera of recipient cytoplasts on development and viability of Canada lynx (*Lynx canadensis*) cloned embryos. *Reproduction, Fertility and Development* 22(1):186 (abstract).

Goodrowe KL, Miller AM, Wildt DE. 1989. *In vitro* fertilization of gonadotropin-stimulated leopard cat (*Felis bengalensis*) follicular oocytes. *Journal of Experimental Zoology* 252(1):89-95.

Goodrowe KL, Wall RJ, O'Brien SJ, Schmidt PM, Wildt DE. 1988. Developmental competence of domestic cat follicular oocytes after fertilization *in vitro*. *Biology of Reproduction* 39(2):355-372.

Gook DA, Edgar DH, Borg J, Archer J, Lutjen PJ, McBain JC. 2003. Oocyte maturation, follicle rupture and luteinization in human cryopreserved ovarian tissue following xenografting. *Human Reproduction* 18(9):1772-1781.

Gook DA, McCully BA, Edgar DH, McBain JC. 2001. Development of antral follicles in human cryopreserved ovarian tissue following xenografting. *Human Reproduction* 16(3):417-422.

Gook DA, Osborn SM, Johnston WIH. 1993. Cryopreservation of mouse and human oocytes using 1, 2-propanediol and the configuration of the meiotic spindle. *Human Reproduction* 8(7):1101-1109.

Göritz F, Neubauer K, Naidenko SV, Fickel J, Jewgenow K. 2006. Investigations on reproductive physiology in the male Eurasian lynx (*Lynx lynx*). *Theriogenology* 66(6-7):1751-1754.

Gosden RG, Baird DT, Wade JC, Webb R. 1994a. Restoration of fertility to oophorectomized sheep by ovarian autografts stored at -196 degrees C. *Human Reproduction* 9(4):597-603.

Gosden RG, Boulton MI, Grant K, Webb R. 1994b. Follicular development from ovarian xenografts in SCID mice. *Journal of Reproduction and Fertility* 101(3):619-623.

Goto K, Kinoshita A, Takuma Y, Ogawa K. 1991. Birth of calves after the transfers of oocytes fertilized by sperm injection. *Theriogenology* 35(1):205 (abstract).

Groves CP, Fernando P, Robovský J. 2010. The sixth rhino: A taxonomic re-assessment of the critically endangered Northern white rhinoceros. *PLoS One* 5(4):e9703.

Guenther JF, Seki S, Kleinhans FW, Edashige K, Roberts DM, Mazur P. 2006. Extra- and intra-cellular ice formation in Stage I and II Xenopus laevis oocytes. *Cryobiology* 52(3):401-416.

Guidetti R, ouml, nsson KI. 2002. Long-term anhydrobiotic survival in semi-terrestrial micrometazoans. *Journal of Zoology* 257(2):181-187.

Gunasena KT, Lakey JRT, Villines PM, Bush M, Raath C, Critser ES, McGann LE, Critser JK. 1998. Antral follicles develop in xenografted cryopreserved african elephant (*Loxodonta africana*) ovarian tissue. *Animal Reproduction Science* 53(1-4):265-275.

Gupta MK, Uhm SJ, Lee HT. 2010. Effect of vitrification and beta-mercaptoethanol on reactive oxygen species activity and *in vitro* development of oocytes vitrified before or after *in vitro* fertilization. *Fertility and Sterility* 93(8):2602-2607.

Gvakharia M, Adamson GD. 2001. A method of successful cryopreservation of small numbers of human spermatozoa. *Fertility and Sterility* 76(3, Suppl. 1):S101 (Abstract).

Hagedorn M, Hsu E, Kleinhans FW, Wildt DE. 1997a. New approaches for studying the permeability of fish embryos: Toward successful cryopreservation. *Cryobiology* 34(4):335-347.

Hagedorn M, Hsu EW, Pilatus U, Wildt DE, Rall WR, Blackband SJ. 1996. Magnetic resonance microscopy and spectroscopy reveal kinetics of cryoprotectant permeation in a multicompartmental biological system. *Proceedings of the National Academy of Science of the Uunited States of America* 93(15):7454-7459.

Hagedorn M, Kleinhans FW, Freitas R, Liu J, Hsu EW, Wildt DE, Rall WF. 1997b. Water distribution and permeability of zebrafish embryos, *Brachydanio rerio*. *Journal of Experimental Zoology* 278(6):356-371.

Hagedorn M, Kleinhans FW, Wildt DE, Rall WF. 1997c. Chill sensitivity and cryoprotectant permeability of dechorionated zebrafish embryos, *Brachydanio rerio*. *Cryobiology* 34(3):251-263.

Hagedorn M, Lance SL, Fonseca DM, Kleinhans FW, Artimov D, Fleischer R, Hoque AT, Hamilton MB, Pukazhenthi BS. 2002. Altering fish embryos with aquaporin-3: an essential step toward successful cryopreservation. *Biology of Reproduction* 67(3):961-966.

Hamawaki A, Kuwayama M, Hamano S. 1999. Minimum volume cooling method for bovine blastocyst vitrification. *Theriogenology* 51(1):165 (abstract).

Hammerstedt RH, Graham JK, Nolan JP. 1990. Cryopreservation of mammalian sperm: what we ask them to survive. *Journal of Andrology* 11(1):73-88.

Hassa H, Gurer F, Yildirim A, Can C, Sahinturk V, Tekin B. 2006. A new protection solution for freezing small numbers of sperm inside empty zona pellucida: Osmangazi-Turk Solution. *Cell Preservation Technology* 4(3):199-208.

Hayakawa H, Yamazaki T, Oshi M, Hoshino M, Dochi O, Koyama H. 2007. Cryopreservation of conventional and sex-sorted bull sperm using a directional freezing method. *Reproduction, Fertility and Development* 19(1):176-177 (abstract).

Hayashi S, Kobayashi K, Mizuno J, Saitoh K, Hirano S. 1989. Birth of piglets from frozen embryos. *The Veterinary Record* 125(2):43-44.

Hearn JP, Summers PM. 1986. Experimental manipulation of embryo implantation in the marmoset monkey and exotic equids. *Theriogenology* 25(1):3-11.

Hermes R, Arav A, Saragusty J, Göritz F, Pettit M, Blottner S, Flach E, Eshkar G, Boardman W, Hildebrandt TB. Cryopreservation of Asian elephant spermatozoa using directional freezing; 2003 4.10.-10.10.2003; Minneapolis, MN, USA. p 264 (abstract).

Hermes R, Behr B, Hildebrandt TB, Blottner S, Sieg B, Frenzel A, Knieriem A, Saragusty J, Rath D. 2009a. Sperm sex-sorting in the Asian elephant (*Elephas maximus*). *Animal Reproduction Science* 112(3-4):390-396.

Hermes R, Göritz F, Portas TJ, Bryant BR, Kelly JM, Maclellan LJ, Keeley T, Schwarzenberger F, Walzer C, Schnorrenberg A and others. 2009b.

Ovarian superstimulation, transrectal ultrasound-guided oocyte recovery, and IVF in rhinoceros. *Theriogenology* 72(7):959-968.

Hermes R, Göritz F, Saragusty J, Sos E, Molnar V, Reid CE, Schwarzenberger F, Hildebrandt TB. 2009c. First successful artificial insemination with frozen-thawed semen in rhinoceros. *Theriogenology* 71(3):393-399.

Herrera C, Conde P, Donaldson M, Quintans C, Cortvrindt R, de Matos DG. 2002. Bovine follicular development up to antral stages after frozen-thawed ovarian tissue transplantation into nude mice. *Theriogenology* 57(1):608 (anstract).

Herrler A, Eisner S, Bach V, Weissenborn U, Beier HM. 2006. Cryopreservation of spermatozoa in alginic acid capsules. *Fertility and Sterility* 85(1):208-213.

Hewitson L, Martinovich C, Simerly C, Takahashi D, Schatten G. 2002. Rhesus offspring produced by intracytoplasmic injection of testicular sperm and elongated spermatids. *Fertility and Sterility* 77(4):794-801.

Higaki S, Eto Y, Kawakami Y, Yamaha E, Kagawa N, Kuwayama M, Nagano M, Katagiri S, Takahashi Y. 2010. Production of fertile zebrafish (*Danio rerio*) possessing germ cells (gametes) originated from primordial germ cells recovered from vitrified embryos. *Reproduction* 139(4):733-740.

Hildebrandt TB, Hermes R, Jewgenow K, Göritz F. 2000a. Ultrasonography as an important tool for the development and application of reproductive technologies in non-domestic species. *Theriogenology* 53(1):73-84.

Hildebrandt TB, Hermes R, Pratt NC, Fritsch G, Blottner S, Schmitt DL, Ratanakorn P, Brown JL, Rietschel W, Göritz F. 2000b. Ultrasonography of the urogenital tract in elephants (*Loxodonta africana* and *Elephas maximus*): an important tool for assessing male reproductive function. *Zoo Biology* 19(5):333-345.

Hildebrandt TB, Roellig K, Goeritz F, Fassbender M, Krieg R, Blottner S, Behr B, Hermes R. 2009. Artificial insemination of captive European brown hares (*Lepus europaeus* PALLAS, 1778) with fresh and cryopreserved semen derived from free-ranging males. *Theriogenology* 72(8):1065-1072.

Hirabayashi M, Kato M, Hochi S. 2008. Factors affecting full-term development of rat oocytes microinjected with fresh or cryopreserved round spermatids. *Experimental Animals* 57(4):401-405.

Hirabayashi M, Kato M, Ito J, Hochi S. 2005. Viable rat offspring derived from oocytes intracytoplasmically injected with freeze-dried sperm heads. *Zygote* 13(1):79-85.

Hochi S, Watanabe K, Kato M, Hirabayashi M. 2008. Live rats resulting from injection of oocytes with spermatozoa freeze-dried and stored for one year. *Molecular Reproduction and Development* 75(5):890-894.

Holt WV. 2001. Germplasm Cryopreservation in Elephants and Wild Ungulates. In: Watson PF, Holt WV, editors. *Cryobanking the Genetic Resource: Wildlife Conservation for the Future.* New York: Taylor & Francis. p 317-348.

Holt WV, Bennett PM, Volobouev V, Watwon PF. 1996. Genetic resource banks in wildlife conservation. *Journal of Zoology* 238(3):531-544.

Holt WV, Pickard AR. 1999. Role of reproductive technology and genetic resource banks in animal conservation. *Reviews of Reproduction* 4:143-150.

Honaramooz A, Behboodi E, Blash S, Megee SO, Dobrinski I. 2003a. Germ cell transplantation in goats. *Molecular Reproduction and Development* 64(4):422-428.

Honaramooz A, Behboodi E, Megee SO, Overton SA, Galantino-Homer H, Echelard Y, Dobrinski I. 2003b. Fertility and germline transmission of donor haplotype following germ cell transplantation in immunocompetent goats. *Biology of Reproduction* 69(4):1260-1264.

Honaramooz A, Li MW, Penedo MC, Meyers S, Dobrinski I. 2004. Accelerated maturation of primate testis by xenografting into mice. *Biology of Reproduction* 70(5):1500-1503.

Honaramooz A, Megee SO, Dobrinski I. 2002. Germ cell transplantation in pigs. *Biology of Reproduction* 66(1):21-28.

Hong Y, Liu T, Zhao H, Xu H, Wang W, Liu R, Chen T, Deng J, Gui J. 2004. Establishment of a normal medakafish spermatogonial cell line capable of sperm production *in vitro*. *Proceedings of the National Academy of Science of the Uunited States of America* 101(21):8011-8016.

Hosu BG, Mullen SF, Critser JK, Forgacs G. 2008. Reversible disassembly of the actin cytoskeleton improves the survival rate and developmental competence of cryopreserved mouse oocytes. *PLoS One* 3(7):e2787.

Hovatta O, Foudila T, Siegberg R, Johansson K, von Smitten K, Reima I. 1996. Pregnancy resulting from intracytoplasmic injection of spermatozoa from a frozen-thawed testicular biopsy specimen. *Human Reproduction* 11(11):2472-2473.

Howell-Stephens JA, Brown J, Santymire R, Bernier D, Mulkerin D. Using giving-up densities and adrenocortical activity to determine the state of Southern three-banded armadillos (*Tolypeutes matacus*) housed in North America; 2009 9-14.8.2009; Mendoza, Argentina. p 190 (abstract).

Hsieh Y-Y, Tsai H-D, Chang C-C, Lo H-Y. 2000. Cryopreservation of human spermatozoa within human or mouse empty zona pellucidae. *Fertility and Sterility* 73(4):694-698.

Hu JCY, Neri QV, Kocent J, Rosenwaks Z, Palermo GD. 2010. Cryopreservation of individual spermatozoa for men with compromised spermatogenesis. *Fertility and Sterility* 94(4, Supplement 1):S113 (abstract).

Imhof M, Hofstetter G, Bergmeister H, Rudas M, Kain R, Lipovac M, Huber J. 2004. Cryopreservation of a whole ovary as a strategy for restoring ovarian function. *Journal of Assisted Reproduction and Genetics* 21(12):459-465.

Isachenko E, Isachenko V, Katkov II, Dessole S, Nawroth F. 2003a. Vitrification of mammalian spermatozoa in the absence of cryoprotectants: from past practical difficulties to present success. *Reproductive BioMedicine Online* 6(2):191-200.

Isachenko E, Isachenko V, Weiss J, Kreienberg R, Katkov I, Schulz M, Lulat A, Risopatron J, Sanchez R. 2008. Acrosomal status and mitochondrial activity of human spermatozoa vitrified with sucrose. *Reproduction* 136(2):167-173.

Isachenko V, Alabart JL, Nawroth F, Isachenko E, Vajta G, Folch J. 2001. The open pulled straw vitrification of ovine GV-oocytes: positive effect of rapid cooling or rapid thawing or both? *Cryo Letters* 22(3):157-162.

Isachenko V, Folch J, Isachenko E, Nawroth F, Krivokharchenko A, Vajta G, Dattena M, Alabart JL. 2003b. Double vitrification of rat embryos at different developmental stages using an identical protocol. *Theriogenology* 60(3):445-452.

Isachenko V, Isachenko E, Katkov II, Montag M, Dessole S, Nawroth F, van der Ven H. 2004. Cryoprotectant-free cryopreservation of human spermatozoa by vitrification and freezing in vapor: Effect on motility, DNA integrity, and fertilization ability. *Biology of Reproduction* 71(4):1167-1173.

Isachenko V, Lapidus I, Isachenko E, Krivokharchenko A, Kreienberg R, Woriedh M, Bader M, Weiss JM. 2009. Human ovarian tissue vitrification versus conventional freezing: morphological, endocrinological, and molecular biological evaluation. *Reproduction* 138(2):319-327.

Isaev DA, Zaletov SY, Zaeva VV, Zakharova EE, Shafei RA, Krivokharchenko IS. 2007. Artificial microcontainers for cryopreservation of solitary spermatozoa. *Human Reproduction* 22(suppl 1):i154-i155 (abstract).

IUCN. 1987. *IUCN Policy Statement on Captive Breeding.* Gland, Switzerland.

IUCN. 2004. IUCN Red List of Threatened Species. Gland, Switzerland: International Union for the Conservation of Nature. http://www.iucnredlist.org, Accessed on August 6, 2007.

Iwayama H, Hochi S, Kato M, Hirabayashi M, Kuwayama M, Ishikawa H, Ohsumi S, Fukui Y. 2005. Effects of cryodevice type and donors' sexual maturity on vitrification of minke whale (*Balaenoptera bonaerensis*) oocytes at germinal vesicle stage. *Zygote* 12(4):333-338.

Izadyar F, Den Ouden K, Stout TA, Stout J, Coret J, Lankveld DP, Spoormakers TJ, Colenbrander B, Oldenbroek JK, Van der Ploeg KD and others. 2003. Autologous and homologous transplantation of bovine spermatogonial stem cells. *Reproduction* 126(6):765-774.

Jang G, Kim MK, Oh HJ, Hossein MS, Fibrianto YH, Hong SG, Park JE, Kim JJ, Kim HJ, Kang SK and others. 2007. Birth of viable female dogs produced by somatic cell nuclear transfer. *Theriogenology* 67(5):941-947.

Janik M, Kleinhans FW, Hagedorn M. 2000. Overcoming a permeability barrier by microinjecting cryoprotectants into zebrafish embryos (*Brachydanio rerio*). *Cryobiology* 41(1):25-34.

Jennings TA. 2002. Lyophilization: Introduction and Basic Principles. Boca Raton, Florida: CRC Press. xvii, 646 p.

Jewgenow K, Blottner S, Lengwinat T, Meyer HH. 1997. New methods for gamete rescue from gonads of nondomestic felids. *Journal of Reproduction and Fertility, Supplement* 51:33-39.

Jewgenow K, Paris MC. 2006. Preservation of female germ cells from ovaries of cat species. *Theriogenology* 66(1):93-100.

Jewgenow K, Penfold LM, Meyer HHD, Wildt DE. 1998. Viability of small preantral ovarian follicles from domestic cats after cryoprotectant exposure and cryopreservation. *Journal of Reproduction and Fertility* 112(1):39-47.

Jin B, Mochida K, Ogura A, Hotta E, Kobayashi Y, Ito K, Egawa G, Seki S, Honda H, Edashige K and others. 2010. Equilibrium vitrification of mouse embryos. *Biology of Reproduction* 82(2):444-450.

Johnson LA. 1988. Flow cytometric determination of sperm sex ratio in semen purportedly enriched for X- or Y-bearing sperm. *Theriogenology* 29(1):265.

Johnson LA, Flook JP, Look MV, Pinkel D. 1987. Flow sorting of X and Y chromosome-bearing spermatozoa into two populations. *Gamete Research* 16(1):1-9.

Johnston LA, Donoghue AM, O'Brien SJ, Wildt DE. 1991. Rescue and maturation *in vitro* of follicular oocytes collected from nondomestic felid species. *Biology of Reproduction* 45(6):898-906.

Johnston LA, Lacy RC. 1995. Genome resource banking for species conservation: Selection of sperm donors. *Cryobiology* 32(1):68-77.

Johnston SD, McGowan MR, Carrick FN, Tribe A. 1993. Preliminary investigations into the feasibility of freezing koala (*Phascolarctos cinereus*) semen. *Australian Veterinary Journal* 70(11):424-425.

Johnston SD, O'Callaghan P, Nilsson K, Tzipori G, Curlewis JD. 2004. Semen-induced luteal phase and identification of a LH surge in the koala (*Phascolarctos cinereus*). *Reproduction* 128(5):629-634.

Just A, Gruber I, Wöber M, Lahodny J, Obruca A, Strohmer H. 2004. Novel method for the cryopreservation of testicular sperm and ejaculated spermatozoa from patients with severe oligospermia: A pilot study. *Fertility and Sterility* 82(2):445-447.

Kagawa N, Kuwayama M, Silber SJ, Takehara Y, Kato K, Kato O. 2007. Successful ovarian tissue vitrification in mouse, bovine and human. *Fertility and Sterility* 88(Supplement 1):S275-S276 (abstract).

Kagawa N, Silber S, Kuwayama M. 2009. Successful vitrification of bovine and human ovarian tissue. *Reproductive Biomedicine Online* 18(4):568-577.

Kalicharan D, Jongebloed WL, Rawson DM, Zhang T. 1998. Variations in fixation techniques for field emission SEM and TEM of zebrafish (*Branchydanio rerio*) embryo inner and outer membranes. *Journal of Electron Microscopy* (Tokyo) 47(6):645-658.

Kaneko T, Nakagata N. 2005. Relation between storage temperature and fertilizing ability of freeze-dried mouse spermatozoa. *Comparative Medicine* 55(2):140-144.

Kaneko T, Whittingham DG, Yanagimachi R. 2003. Effect of pH value of freeze-drying solution on the chromosome integrity and developmental ability of mouse spermatozoa. *Biology of Reproduction* 68(1):136-139.

Kanno H, Saito K, Ogawa T, Takeda M, Iwasaki A, Kinoshita Y. 1998. Viability and function of human sperm in electrolyte-free cold preservation. *Fertility and Sterility* 69(1):127-131.

Kanno H, Speedy RJ, Angell CA. 1975. Supercooling of Water to -92°C Under Pressure. *Science* 189(4206):880-881.

Kasai M. 2002. Advances in the cryopreservation of mammalian oocytes and embryos: Development of ultrarapid vitrification. *Reproductive Medicine and Biology* 1(1):1-9.

Kasai M, Iritani A, Chang MC. 1979. Fertilization *in vitro* of rat ovarian oocytes after freezing and thawing. *Biology of Reproduction* 21(4):839-844.

Kasiraj R, Misra AK, Mutha Rao M, Jaiswal RS, Rangareddi NS. 1993. Successful culmination of pregnancy and live birth following the transfer of frozen-thawed buffalo embryos. *Theriogenology* 39(5):1187-1192.

Katayose H, Matsuda J, Yanagimachi R. 1992. The ability of dehydrated hamster and human sperm nuclei to develop into pronuclei. *Biology of Reproduction* 47(2):277-284.

Kato H, Anzai M, Mitani T, Morita M, Nishiyama Y, Nakao A, Kondo K, Lazarev PA, Ohtani T, Shibata Y and others. 2009. Recovery of cell nuclei from 15,000-year-old mammoth tissue and injection into mouse enucleated matured oocytes. *Reproduction, Fertility and Development* 22(1):189 (abstract).

Kawase Y, Araya H, Kamada N, Jishage K, Suzuki H. 2005. Possibility of long-term preservation of freeze-dried mouse spermatozoa. *Biology of Reproduction* 72(3):568-573.

Keros V, Xella S, Hultenby K, Pettersson K, Sheikhi M, Volpe A, Hreinsson J, Hovatta O. 2009. Vitrification versus controlled-rate freezing in cryopreservation of human ovarian tissue. *Human Reproduction* 24(7):1670-1683.

Keskintepe L, Pacholczyk G, Machnicka A, Norris K, Curuk MA, Khan I, Brackett BG. 2002. Bovine blastocyst development from oocytes injected with freeze-dried spermatozoa. *Biology of Reproduction* 67(2):409-415.

Kim MK, Jang G, Oh HJ, Yuda F, Kim HJ, Hwang WS, Hossein MS, Kim JJ, Shin NS, Kang SK and others. 2007. Endangered wolves cloned from adult somatic cells. *Cloning and Stem Cells* 9(1):130-137.

Kim SS, Kang HG, Kim NH, Lee HC, Lee HH. 2005. Assessment of the integrity of human oocytes retrieved from cryopreserved ovarian tissue after xenotransplantation. *Human Reproduction* 20(9):2502-2508.

Kimura Y, Yanagimachi R. 1995. Mouse oocytes injected with testicular spermatozoa or round spermatids can develop into normal offspring. *Development* 121(8):2397-2405.

Kleinhans FW, Guenther JF, Roberts DM, Mazur P. 2006. Analysis of intracellular ice nucleation in Xenopus oocytes by differential scanning calorimetry. *Cryobiology* 52(1):128-138.

Koh LP, Dunn RR, Sodhi NS, Colwell RK, Proctor HC, Smith VS. 2004. Species coextinctions and the biodiversity crisis. *Science* 305:1632-1634.

Kono T, Suzuki O, Tsunoda Y. 1988. Cryopreservation of rat blastocysts by vitrification. *Cryobiology* 25(2):170-173.

Koscinski I, Wittemer C, Lefebvre-Khalil V, Marcelli F, Defossez A, Rigot JM. 2007. Optimal management of extreme oligozoospermia by an appropriate cryopreservation programme. *Human Reproduction* 22(10):2679-2684.

Kouba AJ, Vance CK. 2009. Applied reproductive technologies and genetic resource banking for amphibian conservation. *Reproduction, Fertility and Development* 21(6):719-737.

Kramer L, Dresser BL, Pope CE, Dalhausen RD, Baker RD. The non-surgical transfer of frozen-thawed eland (*Tragelaphus oryx*) embryos. In: Fowler ME, editor; 1983 October 24-27, 1983; Tampa, Florida, USA. p 104-105.

Kretzschmar P, Ganslosser U, Dehnhard M. 2004. Relationship between androgens, environmental factors and reproductive behavior in male white rhinoceros (*Ceratotherium simum simum*). *Hormones and Behavior* 45(1):1-9.

Krisher RL. 2004. The effect of oocyte quality on development. *Journal of Animal Science* 82(13_suppl):E14-E23.

Kusakabe H, Szczygiel MA, Whittingham DG, Yanagimachi R. 2001. Maintenance of genetic integrity in frozen and freeze-dried mouse spermatozoa. *Proceedings of the National Academy of Sciences of the United States of America* 98(24):13501-13506.

Kuwayama M, Kato O. 2000. All-round vitrification method for human oocytes and embryos. *Journal of Assisted Reproduction and Genetics* 17(8):477 (abstract).

Kuwayama M, Vajta G, Kato O, Leibo SP. 2005. Highly efficient vitrification method for cryopreservation of human oocytes. *Reproduction Biomedicine Online* 11(3):300-308.

Kwon IK, Park KE, Niwa K. 2004. Activation, pronuclear formation, and development *in vitro* of pig oocytes following intracytoplasmic injection of freeze-dried spermatozoa. *Biology of Reproduction* 71(5):1430-1436.

Landa CA, Almquist JO. 1979. Effect of freezing large numbers of straws of bovine spermatozoa in an automatic freezer on post-thaw motility and acrosomal retention. *Journal of Animal Science* 49(5):1190-1194.

Lane M, Forest KT, Lyons EA, Bavister BD. 1999a. Live births following vitrification of hamster embryos using a novel containerless technique. *Theriogenology* 51(1):167 (abstract).

Lane M, Schoolcraft WB, Gardner DK, Phil D. 1999b. Vitrification of mouse and human blastocysts using a novel cryoloop container-less technique. *Fertility and Sterility* 72(6):1073-1078.

Langlais J, Roberts KD. 1985. A molecular membrane model of sperm capacitation and the acrosome reaction of mammalian spermatozoa. *Gamete Research* 12:183-224.

Lanza RP, Cibelli JB, Diaz F, Moraes CT, Farin PW, Farin CE, Hammer CJ, West MD, Damiani P. 2000. Cloning of an endangered species (*Bos gaurus*) using interspecies nuclear transfer. *Cloning* 2(2):79-90.

Lanzendorf SE, Zelinski-Wooten MB, Stouffer RL, Wolf DP. 1990. Maturity at collection and the developmental potential of rhesus monkey oocytes. *Biology of Reproduction* 42(4):703-711.

Larman MG, Gardner DK. 2010. Vitrifying mouse oocytes and embryos with super-cooled air. *Human Reproduction* 25(Suppl. 1):i265 (abstract).

Larman MG, Gardner DK. 2011. Vitrification of mouse embryos with super-cooled air. *Fertility and Sterility* 95(4):1462-1466.

Larson EV, Graham EF. 1976. Freeze-drying of spermatozoa. *Developments in Biological Standartization* 36:343-348.

Lattanzi M, Santos C, Chaves G, Miragaya M, Capdevielle E, Judith E, Agüero A, Baranao L. 2002. Cryopreservation of llama (*Lama glama*) embryos by slow freezing and vitrification. *Theriogenology* 57(1):585 (abstract).

Lee B, Wirtu GG, Damiani P, Pope E, Dresser BL, Hwang W, Bavister BD. 2003. Blastocyst development after intergeneric nuclear transfer of mountain bongo antelope somatic cells into bovine oocytes. *Cloning and Stem Cells* 5(1):25-33.

Lee DM, Yeoman RR, Battaglia DE, Stouffer RL, Zelinski-Wooten MB, Fanton JW, Wolf DP. 2004. Live birth after ovarian tissue transplant. *Nature* 428(6979):137-138.

Lee DR, Yang YH, Eum JH, Seo JS, Ko JJ, Chung HM, Yoon TK. 2007. Effect of using slush nitrogen (SN2) on development of microsurgically manipulated vitrified/warmed mouse embryos. *Human Reproduction* 22(9):2509-2514.

Lee H-J, Elmoazzen H, Wright D, Biggers J, Rueda BR, Heo YS, Toner M, Toth TL. 2010. Ultra-rapid vitrification of mouse oocytes in low cryoprotectant concentrations. *Reproductive Biomedicine Online* 20(2):201-208.

Leibo SP. 1980. Water permeability and its activation energy of fertilized and unfertilized mouse ova. *Journal of Membrane Biology* 53(3):179-188.

Leidy C, Gousset K, Ricker J, Wolkers W, Tsvetkova N, Tablin F, Crowe J. 2004. Lipid phase behavior and stabilization of domains in membranes of platelets. *Cell Biochemistry and Biophysics* 40(2):123-148.

Leslie SB, Teter SA, Crowe LM, Crowe JH. 1994. Trehalose lowers membrane phase transitions in dry yeast cells. *Biochimica et Biophysica Acta (BBA) - Biomembranes* 1192(1):7-13.

Li QY, Hou J, Chen YF, An XR. 2010. Full-term development of rabbit embryos produced by ICSI with sperm frozen in liquid nitrogen without cryoprotectants. *Reproduction in Domestic Animals* 45(4):717-722.

Li YB, Zhou CQ, Yang GF, Wang Q, Dong Y. 2007. Modified vitrification method for cryopreservation of human ovarian tissues. *Chinese Medical Journal* 120(2):110-114.

Li Z, Sun X, Chen J, Liu X, Wisely SM, Zhou Q, Renard JP, Leno GH, Engelhardt JF. 2006. Cloned ferrets produced by somatic cell nuclear transfer. *Developmental Biology* 293(2):439-448.

Liebermann J, Tucker M, Graham J, Han T, Davis A, Levy M. 2002. Blastocyst development after vitrification of multipronuclear zygotes using the felxipet denuding pipette. *Reproduction Biomedicine Online* 4(2):146-150.

Lim JJ, Shin TE, Song S-H, Bak CW, Yoon TK, Lee DR. 2010. Effect of liquid nitrogen vapor storage on the motility, viability, morphology, deoxyribonucleic acid integrity, and mitochondrial potential of frozen-thawed human spermatozoa. *Fertility and Sterility* 94(7):2736-2741.

Lin T-T, Pitt RE, Steponkus PL. 1989. Osmometric behavior of *Drosophila melanogaster* embryos. *Cryobiology* 26(5):453-471.

Lindeberg H, Aalto J, Amstislavsky S, Piltti K, Järvinen M, Valtonen M. 2003. Surgical recovery and successful surgical transfer of conventionally frozen-thawed embryos in the farmed European polecat (*Mustela putorius*). *Theriogenology* 60(8):1515-1525.

Liu J, Van der Elst J, Van den Broecke R, Dhont M. 2001. Live offspring by *in vitro* fertilization of oocytes from cryopreserved primordial mouse follicles after sequential *in vivo* transplantation and *in vitro* maturation. *Biology of Reproduction* 64(1):171-178.

Liu J, Van der Elst J, Van den Broecke R, Dhont M. 2002. Early massive follicle loss and apoptosis in heterotopically grafted newborn mouse ovaries. *Human Reproduction* 17(3):605-611.

Liu J, Van Der Elst J, Van Den Broecke R, Dumortier F, Dhont M. 2000. Maturation of mouse primordial follicles by combination of grafting and *in vitro* culture. *Biology of Reproduction* 62(5):1218-1223.

Liu JL, Kusakabe H, Chang CC, Suzuki H, Schmidt DW, Julian M, Pfeffer R, Bormann CL, Tian XC, Yanagimachi R and others. 2004. Freeze-dried sperm fertilization leads to full-term development in rabbits. *Biology of Reproduction* 70(6):1776-1781.

Liu L, Wood GA, Morikawa L, Ayearst R, Fleming C, McKerlie C. 2008. Restoration of fertility by orthotopic transplantation of frozen adult mouse ovaries. *Human Reproduction* 23(1):122-128.

Living Planet Report 2008. 2008. Gland, Switzerland: *World Wildlife Fund* (WWF).

Locatelli Y, Vallet J-C, Baril G, Touzé J-L, Hendricks A, Legendre X, Verdier M, Mermillod P. 2008. Successful interspecific pregnancy after transfer of *in vitro* produced sika deer (*Cervus nippon nippon*) embryo in red deer (*Cervus elaphus hippelaphus*) surrogate hind. *Reproduction, Fertility and Development* 20(1):160-161 (abstract).

Loi P, Clinton M, Barboni B, Fulka J, Jr., Cappai P, Feil R, Moor RM, Ptak G. 2002. Nuclei of nonviable ovine somatic cells develop into lambs after nuclear transplantation. *Biology of Reproduction* 67(1):126-132.

Loi P, Matsukawa K, Ptak G, Clinton M, Fulka Jr. J, Natan Y, Arav A. 2008a. Freeze-dried somatic cells direct embryonic development after nuclear transfer. *PLoS One* 3(8):e2978.

Loi P, Matsukawa K, Ptak G, Nathan Y, Fulka J, Jr., Arav A. 2008b. Nuclear transfer of freeze dried somatic cells into enucleated sheep oocytes. *Reproduction in Domestic Animals* 43(Suppl. 2):417-422.

Loi P, Ptak G, Barboni B, Fulka J, Jr., Cappai P, Clinton M. 2001. Genetic rescue of an endangered mammal by cross-species nuclear transfer using post-mortem somatic cells. *Nature Biotechnology* 19(10):962-964.

Loskutoff NM, Bartels P, Meintjes M, Godke RA, Schiewe MC. 1995. Assisted reproductive technology in nondomestic ungulates: A model approach to preserving and managing genetic diversity. *Theriogenology* 43(1):3-12.

Lu F, Shi D, Wei J, Yang S, Wei Y. 2005. Development of embryos reconstructed by interspecies nuclear transfer of adult fibroblasts between buffalo (*Bubalus bubalis*) and cattle (*Bos indicus*). *Theriogenology* 64(6):1309-1319.

Luvoni GC. 2000. Current progress on assisted reproduction in dogs and cats: *in vitro* embryo production. *Reproduction, Nutrition, Development* 40(5):505-512.

Luvoni GC. 2006. Gamete cryopreservation in the domestic cat. *Theriogenology* 66(1):101-111.

Luvoni GC, Pellizzari P, Battocchio M. 1997. Effects of slow and ultrarapid freezing on morphology and resumption of meiosis in immature cat oocytes. *Journal of Reproduction and Fertility Supplement* 51:93-98.

Luyet B. 1937. The vitrification of organic colloids and protoplasm. *Biodynamica* 1(29):1-14.

Luyet BJ, Hoddap A. 1938 Revival of frog's sprmatozoa vitrified in liquid air. Proceding of the Meeting of the Society for Experimental Biology, p 433-434.

Magli MC, Lappi M, Ferraretti AP, Capoti A, Ruberti A, Gianaroli L. 2010. Impact of oocyte cryopreservation on embryo development. *Fertility and Sterility* 93(2):510-516.

Mahesh YU, Rao BS, Suman K, Lakshmikantan U, Charan KV, Gibence HRW, Shivaji S. 2011. *In vitro* maturation and fertilization in the nilgai (*Boselaphus tragocamelus*) using oocytes and spermatozoa recovered post-mortem from animals that had died because of foot and mouth disease outbreak. *Reproduction in Domestic Animals* In Press.

Margules CR, Pressey RL. 2000. Systematic conservation planning. *Nature* 405(6783):243-253.

Martin JR, Bromer JG, Sakkas D, Patrizio P. 2010. Live babies born per oocyte retrieved in a subpopulation of oocyte donors with repetitive reproductive success. *Fertility and Sterility* 94(6):2064-2068.

Martino A, Songsasen N, Leibo SP. 1996. Development into blastocysts of bovine oocytes cryopreserved by ultra-rapid cooling. *Biology of Reproduction* 54(5):1059-1069.

Martins CF, Bao SN, Dode MN, Correa GA, Rumpf R. 2007. Effects of freeze-drying on cytology, ultrastructure, DNA fragmentation, and fertilizing ability of bovine sperm. *Theriogenology* 67(8):1307-1315.

Mastromonaco GF, King WA. 2007. Cloning in companion animal, non-domestic and endangered species: can the technology become a practical reality? *Reproduction, Fertility and Development* 19(6):748-761.

Matsumoto H, Jiang JY, Tanaka T, Sasada H, Sato E. 2001. Vitrification of large quantities of immature bovine oocytes using nylon mesh. *Cryobiology* 42(2):139-144.

Mattiske D, Shaw G, Shaw JM. 2002. Influence of donor age on development of gonadal tissue from pouch young of the tammar wallaby, *Macropus eugenii*, after cryopreservation and xenografting into mice. *Reproduction* 123(1):143-153.

Mazur P, Cole KW, Hall JW, Schreuders PD, Mahowald AP. 1992. Cryobiological preservation of Drosophila embryos. *Science* 258(5090):1932-1935.

Mazur P, Leibo SP, Seidel GE, Jr. 2008. Cryopreservation of the germplasm of animals used in biological and medical research: importance, impact, status, and future directions. *Biology of Reproduction* 78(1):2-12.

McDonnell SM. 2001. Oral imipramine and intravenous xylazine for pharmacologically-induced ex copula ejaculation in stallions. *Animal Reproduction Science* 68(3-4):153-159.

McGann LE, Yang H, Walterson M. 1988. Manifestations of cell damage after freezing and thawing. *Cryobiology* 25(3):178-185.

Melville DF, Crichton EG, Paterson-Wimberley T, Johnston SD. 2008. Collection of semen by manual stimulation and ejaculate characteristics of the black flying-fox (*Pteropus alecto*). *Zoo Biology* 27(2):159-164.

Meyers SA. 2006. Dry storage of sperm: applications in primates and domestic animals. *Reproduction, Fertility and Development* 18(1-2):1-5.

Migishima F, Suzuki-Migishima R, Song SY, Kuramochi T, Azuma S, Nishijima M, Yokoyama M. 2003. Successful cryopreservation of mouse ovaries by vitrification. *Biology of Reproduction* 68(3):881-887.

Miller DL, Waldhalm SJ, Leopold BD, Estill C. 2002. Embryo transfer and embryonic capsules in the bobcat (*Lynx rufus*). *Anatomia, Histologia, Embryologia* 31(2):119-125.

Mitre R, Cheminade C, Allaume P, Legrand P, Legrand AB. 2004. Oral intake of shark liver oil modifies lipid composition and improves motility and velocity of boar sperm. *Theriogenology* 62(8):1557-1566.

Mochida K, Wakayama T, Takano K, Noguchi Y, Yamamoto Y, Suzuki O, Matsuda J, Ogura A. 2005. Birth of offspring after transfer of Mongolian gerbil (*Meriones unguiculatus*) embryos cryopreserved by vitrification. *Molecular Reproduction and Development* 70(4):464-470.

Mochida K, Wakayama T, Takano K, Noguchi Y, Yamamoto Y, Suzuki O, Ogura A, Matsuda J. 1999. Successful cryopreservation of mongolian gerbil embryos by vitrification. *Theriogenology* 51(1):171 (abstract).

Mochida K, Yamamoto Y, Noguchi Y, Takano K, Matsuda J, Ogura A. 2000. Survival and subsequent *in vitro* development of hamster embryos after exposure to cryoprotectant solutions. *Journal of Assisted Reproduction and Genetics* 17(3):182-185.

Moehlman PD. 2002. Status and action plan for the African Wild Ass (*Equus africanus*). Gland: IUCN/SSC Equid Specialist Group. 2-10 p.

Moghadam KK, Nett R, Robins JC, Thomas MA, Awadalla SG, Scheiber MD, Williams DB. 2005. The motility of epididymal or testicular spermatozoa does not directly affect IVF/ICSI pregnancy outcomes. *Journal of Andrology* 26(5):619-623.

Moisan AE, Leibo SP, Lynn JW, Gómez MC, Pope CE, Dresser BL, Godke RA. 2005. Embryonic development of felid oocytes injected with freeze-dried or air-dried spermatozoa. *Cryobiology* 51:373 (abstract).

Montag M, Rink K, Dieckmann U, Delacretaz G, van der Ven H. 1999. Laser-assisted cryopreservation of single human spermatozoa in cell-free zona pellucida. *Andrologia* 31(1):49-53.

Moor RM, Dai Y, Lee C, Fulka J, Jr. 1998. Oocyte maturation and embryonic failure. *Human Reproduction Update* 4(3):223-226.

Moore AI, Squires EL, Graham JK. 2005. Adding cholesterol to the stallion sperm plasma membrane improves cryosurvival. *Cryobiology* 51(3):241-249.

Morais RN, Mucciolo RG, Gomes MLF, Lacerda O, Moraes W, Moreira N, Graham LH, Swanson WF, Brown JL. 2002. Seasonal analysis of semen characteristics, serum testosterone and fecal androgens in the ocelot (*Leopardus pardalis*), margay (*L. wiedii*) and tigrina (*L. tigrinus*). *Theriogenology* 57(8):2027-2041.

Morato RG, Conforti VA, Azevedo FC, Jacomo AT, Silveira L, Sana D, Nunes AL, Guimaraes MA, Barnabe RC. 2001. Comparative analyses of semen and endocrine characteristics of free-living versus captive jaguars (*Panthera onca*). *Reproduction* 122(5):745-751.

Morrow CJ, Asher GW, Berg DK, Tervit HR, Pugh PA, McMillan WH, Beaumont S, Hall DRH, Bell ACS. 1994. Embryo transfer in fallow deer (*Dama dama*): Superovulation, embryo recovery and laparoscopic transfer of fresh and cryopreserved embryos. *Theriogenology* 42(4):579-590.

Morton KM, Ruckholdt M, Evans G, Maxwell WMC. 2008. Quantification of the DNA difference, and separation of X- and Y-bearing sperm in alpacas (*Vicugna pacos*). *Reproduction in Domestic Animals* 43(5):638-642.

Murakami M, Otoi T, Karja NW, Wongsrikeao P, Agung B, Suzuki T. 2004. Blastocysts derived from *in vitro*-fertilized cat oocytes after vitrification and dilution with sucrose. *Cryobiology* 48(3):341-348.

Muthukumar K, Mangalaraj AM, Kamath MS, George K. 2008. Blastocyst cryopreservation: vitrification or slow freeze. *Fertility and Sterility* 90(Supplement 1):S426-S427 (abstract).

Nagano M, Avarbock MR, Leonida EB, Brinster CJ, Brinster RL. 1998. Culture of mouse spermatogonial stem cells. *Tissue and Cell* 30(4):389-397.

Nagashima H, Kashiwazaki N, Ashman RJ, Grupen CG, Nottle MB. 1995. Cryopreservation of porcine embryos. *Nature* 374(6521):416.

Nagy ZP, Chang CC, Shapiro DB, Bernal DP, Kort HI, Vajta G. 2009. The efficacy and safety of human oocyte vitrification. *Seminars in Reproductive Medicine* 27(6):450-455.

Naitana S, Bogliolo L, Ledda S, Leoni G, Madau L, Falchi S, Muzzeddu M. 2000. Survival of vitrified mouflon (*Ovis g. musimon*) blastocysts. *Theriogenology* 53(1):340 (abstract).

Naitana S, Ledda S, Loi P, Leoni G, Bogliolo L, Dattena M, Cappai P. 1997. Polyvinyl alcohol as a defined substitute for serum in vitrification and warming solutions to cryopreserve ovine embryos at different stages of development. *Animal Reproduction Science* 48(2-4):247-256.

Nawroth F, Isachenko V, Dessole S, Rahimi G, Farina M, Vargiu N, Mallmann P, Dattena M, Capobianco G, Peters D and others. 2002. Vitrification of human spermatozoa without cryoprotectants. *Cryo Letters* 23(2):93-102.

Nayudu P, Wu J, Michelmann H. 2003. *In Vitro* development of marmoset monkey oocytes by pre-antral follicle culture. *Reproduction in Domestic Animals* 38(2):90-96.

Nel-Themaat L, Gomez MC, Pope CE, Lopez M, Wirtu G, Jenkins JA, Cole A, Dresser BL, Bondioli KR, Godke RA. 2008. Cloned embryos from semen. Part 2: Intergeneric nuclear transfer of semen-derived eland (*Taurotragus oryx*) epithelial cells into bovine oocytes. *Cloning and Stem Cells* 10(1):161-172.

Nowshari MA, Ali SA, Saleem S. 2005. Offspring resulting from transfer of cryopreserved embryos in camel (*Camelus dromedarius*). *Theriogenology* 63(9):2513-2522.

O'Brien JK, Hollinshead FK, Evans KM, Evans G, Maxwell WM. 2004. Flow cytometric sorting of frozen-thawed spermatozoa in sheep and non-human primates. *Reproduction, Fertility and Development* 15(7):367-375.

O'Brien JK, Robeck TR. 2006. Development of sperm sexing and associated assisted reproductive technology for sex preselection of captive bottlenose dolphins (*Tursiops truncatus*). *Reproduction, Fertility and Development* 18(3):319-329.

O'Brien JK, Robeck TR. 2010a. Preservation of beluga (*Delphinapterus leucas*) spermatozoa using a trehalose-based cryodiluent and directional freezing technology. *Reproduction, Fertility and Development* 22(4):653-663.

O'Brien JK, Robeck TR. 2010b. The value of ex situ Cetacean populations in understanding reproductive physiology and developing assisted reproductive technology for ex situ species management and conservation efforts. *International Journal of Comparative Psychology* 23(3):227-248.

O'Brien JK, Roth TL. 2000. Post-coital sperm recovery and cryopreservation in the Sumatran Rhinoceros (*Dicerorhinus sumatrensis*) and application to gamete rescue in the African black rhinoceros (*Diceros bicornis*). *Journal of Reproduction and Fertility* 118(2):263-271.

O'Brien JK, Stojanov T, Crichton EG, Evans KM, Leigh D, Maxwell WM, Evans G, Loskutoff NM. 2005a. Flow cytometric sorting of fresh and frozen-thawed spermatozoa in the western lowland gorilla (*Gorilla gorilla gorilla*). *American Journal of Primatology* 66(4):297-315.

O'Brien JK, Stojanov T, Heffernan SJ, Hollinshead FK, Vogelnest L, Chis Maxwell WM, Evans G. 2005b. Flow cytometric sorting of non-human primate sperm nuclei. *Theriogenology* 63(1):246-259.

Ogonuki N, Mochida K, Miki H, Inoue K, Fray M, Iwaki T, Moriwaki K, Obata Y, Morozumi K, Yanagimachi R and others. 2006. Spermatozoa and spermatids retrieved from frozen reproductive organs or frozen whole bodies of male mice can produce normal offspring. *Proceedings of the National Academy of Sciences of the United States of America* 103(35):13098-13103.

Okazaki T, Abe S, Yoshida S, Shimada M. 2009. Seminal plasma damages sperm during cryopreservation, but its presence during thawing improves semen quality and conception rates in boars with poor post-thaw semen quality. *Theriogenology* 71(3):491-498.

Oktay K, Cil AP, Bang H. 2006. Efficiency of oocyte cryopreservation: a meta-analysis. *Fertility and Sterility* 86(1):70-80.

Onions VJ, Webb R, McNeilly AS, Campbell BK. 2009. Ovarian endocrine profile and long-term vascular patency following heterotopic autotransplantation of cryopreserved whole ovine ovaries. *Human Reproduction* 24(11):2845-2855.

Pan G, Chen Z, Liu X, Li D, Xie Q, Ling F, Fang L. 2001. Isolation and purification of the ovulation-inducing factor from seminal plasma in the bactrian camel (*Camelus bactrianus*). *Theriogenology* 55(9):1863-1879.

Papahadjopoulos D, Poste G, Schaeffer BE. 1973. Fusion of mammalian cells by unilamellar lipid vesicles: Influence of lipid surface charge, fluidity and cholesterol. *Biochimica et Biophysica Acta (BBA) - Biomembranes* 323(1):23-42.

Papis K, Korwin-Kossakowski M, Wenta-Muchalska E. 2009. Comparison of traditional and modified (VitMaster) methods of rabbit embryo vitrification. *Acta Veterinaria Hungarica* 57(3):411-416.

Paris MCJ, Snow M, Cox S-L, Shaw JM. 2004. Xenotransplantation: a tool for reproductive biology and animal conservation? *Theriogenology* 61(2-3):277-291.

Parrott DMV. 1960. The fertility of mice with orthotopic ovarian grafts derived from frozen tissue. *Journal of Reproduction and Fertility* 1(3):230-241.

Pearl M, Arav A. 2000. Chilling sensitivity in zebrafish (*Brachydanio rerio*) oocytes is related to lipid phase transition. *Cryo Letters* 21(3):171-178.

Petyim S, Makemahar O, Kunathikom S, Choavaratana R, Laokirkkiat P, Penparkkul K. 2009. The successful pregnancy and birth of a healthy baby after human blastocyst vitrification using Cryo-E, first case in Siriraj Hospital. *Journal of the Medical Association of Thailand* 92(8):1116-1121.

Piltti K, Lindeberg H, Aalto J, Korhonen H. 2004. Live cubs born after transfer of OPS vitrified-warmed embryos in the farmed European polecat (*Mustela putorius*). *Theriogenology* 61(5):811-820.

Polcz TE, Stronk J, Xiong C, Jones EE, Olive DL, Huszar G. 1998. Optimal utilization of cryopreserved human semen for assisted reproduction: recovery and maintenance of sperm motility and viability. *Journal of Assisted Reproduction and Genetics* 15(8):504-512.

Polejaeva IA, Chen SH, Vaught TD, Page RL, Mullins J, Ball S, Dai Y, Boone J, Walker S, Ayares DL and others. 2000. Cloned pigs produced by nuclear transfer from adult somatic cells. *Nature* 407(6800):86-90.

Poleo GA, Godke RR, Tiersch TR. 2005. Intracytoplasmic sperm injection using cryopreserved, fixed, and freeze-dried sperm in eggs of Nile tilapia. *Marine Biotechnology* (New York, NY) 7(2):104-111.

Polge C, Smith AU, Parkes AS. 1949. Revival of spermatozoa after vitrification and dehydration at low temperatures. *Nature* 164(4172):666.

Pope CE. 2000. Embryo technology in conservation efforts for endangered felids. *Theriogenology* 53(1):163-174.

Pope CE, Dresser BL, Chin NW, Liu JH, Loskutoff NM, Behnke EJ, Brown C, McRae MA, Sinoway CE, Campbell MK and others. 1997a. Birth of a Western Lowland gorilla (*Gorilla gorilla gorilla*) following *in vitro* fertilization and embryo transfer. *American Journal of Primatology* 41(3):247-260.

Pope CE, Gómez MC, Cole A, Dumas C, Dresser BL. 2005. Oocyte recovery, *in vitro* fertilization and embryo transfer in the serval (*Leptailurus serval*). *Reproduction, Fertility and Development* 18(2):223 (abstract).

Pope CE, Gomez MC, Dresser BL. 2006. *In vitro* embryo production and embryo transfer in domestic and non-domestic cats. *Theriogenology* 66(6-7):1518-1524.

Pope CE, Gomez MC, Dresser BL. 2009. *In vitro* embryo production in the clouded leopard (*Neofelis nebulosa*). *Reproduction, Fertility and Development* 22(1):258 (abstract).

Pope CE, Gomez MC, Harris RF, Dresser BL. 2002. Development of *in vitro* matured, *in vitro* fertilized cat embryos following cryopreservation, culture and transfer. *Theriogenology* 57(1):464 (abstract).

Pope CE, Gómez MC, Mikota SK, Dresser BL. 2000. Development of *in vitro* produced African wild cat (*Felis silvestris*) embryos after cryopreservation and transfer into domestic cat recipients. *Biology of Reproduction 62* (Suppl 1):321 (abstract).

Pope CE, McRae MA, Plair BL, Keller GL, Dresser BL. 1994. Successful *in vitro* and *in vivo* development of *in vitro* fertilized two- to four-cell cat embryos following cryopreservation, culture and transfer. *Theriogenology* 42(3):513-525.

Pope CE, McRae MA, Plair BL, Keller GL, Dresser BL. 1997b. *In vitro* and *in vivo* development of embryos produced by *in vitro* maturation and *in vitro* fertilization of cat oocytes. *Journal of Reproduction and Fertility Supplement* 51:69-82.

Pope CE, Pope VZ, Beck LR. 1984. Live birth following cryopreservation and transfer of a baboon embryo. *Fertility and Sterility* 42(1):143-145.

Popelkova M, Turanova Z, Koprdova L, Ostro A, Toporcerova S, Makarevich AV, Chrenek P. 2009. Effect of vitrification technique and assisted hatching on rabbit embryo developmental rate. *Zygote* 17(1):57-61.

Portmann M, Nagy ZP, Behr B. 2010. Evaluation of blastocyst survival following vitrification / warming using two different closed carrier systems. *Human Reproduction* 25(Suppl. 1):i261 (abstract).

Potts M. 1994. Desiccation tolerance of prokaryotes. *Microbiology Reviews* 58(4):755-805.

Potts M, Slaughter SM, Hunneke F-U, Garst JF, Helm RF. 2005. Desiccation tolerance of prokaryotes: Application of principles to human cells. *Integrative and Comparative Biology* 45(5):800-809.

Pribenszky C, Du Y, Molnar M, Harnos A, Vajta G. 2008. Increased stress tolerance of matured pig oocytes after high hydrostatic pressure treatment. *Animal Reproduction Science* 106(1-2):200-207.

Pribenszky C, Molnar M, Cseh S, Solti L. 2005. Improving post-thaw survival of cryopreserved mouse blastocysts by hydrostatic pressure challenge. *Animal Reproduction Science* 87(1-2):143-150.

Pribenszky C, Molnar M, Horvath A, Harnos A, Szenci O. 2006. Hydrostatic pressure induced increase in post-thaw motility of frozen boar spermatozoa. *Reproduction, Fertility and Development* 18(2):162-163 (abstract).

Ptak G, Clinton M, Barboni B, Muzzeddu M, Cappai P, Tischner M, Loi P. 2002. Preservation of the wild European mouflon: the first example of genetic management using a complete program of reproductive biotechnologies. *Biology of Reproduction* 66(3):796-801.

Pukazhenthi B, Laroe D, Crosier A, Bush LM, Spindler R, Pelican KM, Bush M, Howard JG, Wildt DE. 2006. Challenges in cryopreservation of clouded leopard (*Neofelis nebulosa*) spermatozoa. *Theriogenology* 66(6-7):1790-1796.

Pukazhenthi BS, Wildt DE. 2004. Which reproductive technologies are most relevant to studying, managing and conserving wildlife? *Reproduction, Fertility and Development* 16(2):33-46.

Purdy PH, Graham JK. 2004. Effect of cholesterol-loaded cyclodextrin on the cryosurvival of bull sperm. *Cryobiology* 48(1):36-45.

Qi S, Ma A, Xu D, Daloze P, Chen H. 2008. Cryopreservation of vascularized ovary: an evaluation of histology and function in rats. *Microsurgery* 28(5):380-386.

Rahimi G, Isachenko E, Sauer H, Isachenko V, Wartenberg M, Hescheler J, Mallmann P, Nawroth F. 2003. Effect of different vitrification protocols for human ovarian tissue on reactive oxygen species and apoptosis. *Reproduction, Fertility and Development* 15(6):343-349.

Rall WF. 2001. Cryopreservation of mammalian embryos, gametes, and ovarian tissue. In: Wolf DP, Zelinski-Wooten MB, editors. *Assisted Fertilization and Nuclear Transfer in Mammals*. Totowa, NJ: Humana Press Inc. p 173-187.

Rall WF, Fahy GM. 1985. Ice-free cryopreservation of mouse embryos at -196°C by vitrification. *Nature* 313(6003):573-575.

Rawson DM, Zhang T, Kalicharan D, Jongebloed WL. 2000. Field emission scanning electron microscopy and transmission electron microscopy studies of the chorion, plasma membrane and syncytial layers of the gastrula-stage embryo of the zebrafish *Brachydanio rerio*: a consideration of the structural and functional relationships with respect to cryoprotectant penetration. *Aquaculture Research* 31(3):325-336.

Reid CE, Hermes R, Blottner S, Goeritz F, Wibbelt G, Walzer C, Bryant BR, Portas TJ, Streich WJ, Hildebrandt TB. 2009. Split-sample comparison of directional and liquid nitrogen vapour freezing method on post-thaw semen quality in white rhinoceroses (*Ceratotherium simum simum* and *Ceratotherium simum cottoni*). *Theriogenology* 71(2):275-291.

Reid WV, Miller KR. 1989. *Keeping Options Alive: The scientific basis for the conservation of biodiversity*. Washington, DC: World Resources Institute.

Reubinoff BE, Pera MF, Vajta G, Trounson AO. 2001. Effective cryopreservation of human embryonic stem cells by the open pulled straw vitrification method. *Human Reproduction* 16(10):2187-2194.

Revel A, A E, Bor A, Yavin S, Natan Y, Arav A. 2001. Intact sheep ovary cryopreservation and transplantation. *Fertility and Sterility* 76(3, Suppl. 1):S42-S43 (abstract).

Revel A, Elami A, Bor A, Yavin S, Natan Y, Arav A. 2004. Whole sheep ovary cryopreservation and transplantation. *Fertility and Sterility* 82(6):1714-1715.

Revell SG, Pettit MT, Ford TC. 1997. Use of centrifugation over iodixanol to reduce damage when processing stallion sperm for freezing. *Journal of Reproduction and Fertility Abstract series* 19:38 (abstract 92).

Ridha MT, Dukelow WR. 1985. The developmental potential of frozen-thawed hamster preimplantation embryos following embryo transfer: Viability of slowly frozen embryos following slow and rapid thawing. *Animal Reproduction Science* 9(3):253-259.

Riel JM, Huang TT, Ward MA. 2007. Freezing-free preservation of human spermatozoa--a pilot study. *Archives of Andrology* 53(5):275-284.

Risco R, Elmoazzen H, Doughty M, He X, Toner M. 2007. Thermal performance of quartz capillaries for vitrification. *Cryobiology* 55(3):222-229.

Robeck TR, Steinman KJ, Montano GA, Katsumata E, Osborn S, Dalton L, Dunn JL, Schmitt T, Reidarson T, O'Brien JK. 2010. Deep intra-uterine artificial inseminations using cryopreserved spermatozoa in beluga (*Delphinapterus leucas*). *Theriogenology* 74(6):989-1001.

Robles V, Cabrita E, Herraez MP. 2009. Germplasm cryobanking in zebrafish and other aquarium model species. *Zebrafish* 6(3):281-293.

Robles V, Cabrita E, Real M, Álvarez R, Herráez MP. 2003. Vitrification of turbot embryos: preliminary assays. *Cryobiology* 47(1):30-39.

Rockstrom J, Steffen W, Noone K, Persson A, Chapin FS 3rd, Lambin EF, Lenton TM, Scheffer M, Folke C, Schellnhuber HJ and others. 2009. A safe operating space for humanity. *Nature* 461(7263):472-475.

Rodger JC, Giles I, Mate KE. 1992. Unexpected oocyte growth after follicular antrum formation in four marsupial species. *Journal of Reproduction and Fertility* 96(2):755-763.

Rodrigues BA, Rodrigues JL. 2006. Responses of canine oocytes to *in vitro* maturation and *in vitro* fertilization outcome. *Theriogenology* 66(6-7):1667-1672.

Rodriguez-Gil JE. 2006. Mammalian sperm energy resources management and survival during conservation in refrigeration. *Reproduction in Domestic Animals* 41(Suppl. 2):11-20.

Rodriguez-Sosa JR, Foster RA, Hahnel A. 2010. Development of strips of ovine testes after xenografting under the skin of mice and co-transplantation of exogenous spermatogonia with grafts. *Reproduction* 139(1):227-235.

Roelke ME, Martenson JS, O'Brien SJ. 1993. The consequences of demographic reduction and genetic depletion in the endangered Florida panther. *Current Biology* 3(6):340-350.

Roelke-Parker ME, Munson L, Packer C, Kock R, Cleaveland S, Carpenter M, O'Brien SJ, Pospischil A, Hofmann-Lehmann R, Lutz H and others. 1996. A canine distemper virus epidemic in Serengeti lions (*Panthera leo*). *Nature* 379(6564):441-445.

Rofeim O, Brown TA, Gilbert BR. 2001. Effects of serial thaw-refreeze cycles on human sperm motility and viability. *Fertility and Sterility* 75(6):1242-1243.

Rooke JA, Shao CC, Speake BK. 2001. Effects of feeding tuna oil on the lipid composition of pig spermatozoa and *in vitro* characteristics of semen. *Reproduction* 121(2):315-322.

Roth TL, Bush LM, Wildt DE, Weiss RB. 1999. Scimitar-horned oryx (*Oryx dammah*) spermatozoa are functionally competent in a heterologous bovine *in vitro* fertilization system after cryopreservation on dry ice, in a dry shipper, or over liquid nitrogen vapor. *Biology of Reproduction* 60(2):493-8.

Roth TL, Stoops MA, Atkinson MW, Blumer ES, Campbell MK, Cameron KN, Citino SB, Maas AK. 2005. Semen collection in rhinoceroses (*Rhinoceros unicornis, Diceros bicornis, Ceratotherium simum*) by electroejaculation with a uniquely designed probe. *Journal of Zoo and Wildlife Medicine* 36(4):617-627.

Roth TL, Swanson WF, Collins D, Burton M, Garell DM, Wildt DE. 1996. Snow leopard (*Panthera uncia*) spermatozoa are sensitive to alkaline pH, but motility *in vitro* is not influenced by protein or energy supplements. *Journal of Andrology* 17(5):558-566.

Roussel JD, Kellgren HC, Patrick TE. 1964. Bovine semen frozen in liquid nitrogen vapor. *Journal of Dairy Science* 47(12):1403-1406.

Rubei M, Degl'Innocenti S, De Vries PJ, Catone G, Morini G. Directional freezing (Harmony Cryocare - Multi Thermal Gradient 516): *A new tool for equine semen cryopreservation;* 2004; Porto Seguro, Brazil. p 503 (abstract).

Ruffing NA, Steponkus PL, Pitt RE, Parks JE. 1993. Osmometric behavior, hydraulic conductivity, and incidence of intracellular ice formation in bovine oocytes at different developmental stages. *Cryobiology* 30(6):562-580.

Russell PH, Lyaruu VH, Millar JD, Curry MR, Watson PF. 1997. The potential transmission of infectious agents by semen packaging during storage for artificial insemination. *Animal Reproduction Science* 47(4):337-342.

Ruttimann J. 2006. Doomsday food store takes pole position. *Nature* 441(7096):912-913.

Saacke RG, Almquist JO. 1961. Freeze-drying of Bovine Spermatozoa. *Nature* 192(4806):995-996.

Saito K, Kinoshita Y, Kanno H, Iwasaki A, Hosaka M. 1996. A new method of the electrolyte-free long-term preservation of human sperm at 4 degrees C. *Fertility and Sterility* 65(6):1210-1213.

Salehnia M, Moghadam EA, Velojerdi MR. 2002. Ultrastructure of follicles after vitrification of mouse ovarian tissue. *Fertility and Sterility* 78(3):644-645.

Salle B, Demirci B, Franck M, Berthollet C, Lornage J. 2003. Long-term follow-up of cryopreserved hemi-ovary autografts in ewes: pregnancies, births, and histologic assessment. *Fertility and Sterility* 80(1):172-177.

Sanchez-Partida LG, Simerly CR, Ramalho-Santos J. 2008. Freeze-dried primate sperm retains early reproductive potential after intracytoplasmic sperm injection. *Fertility and Sterility* 89(3):742-745.

Sánchez-Serrano M, Crespo J, Mirabet V, Cobo AC, Escribá M-J, Simón C, Pellicer A. 2010. Twins born after transplantation of ovarian cortical tissue and oocyte vitrification. *Fertility and Sterility* 93(1):268.e11-268.e13.

Santiago-Moreno J, Toledano-Diaz A, Pulido-Pastor A, Gomez-Brunet A, Lopez-Sebastian A. 2006. Birth of live Spanish ibex (*Capra pyrenaica hispanica*) derived from artificial insemination with epididymal spermatozoa retrieved after death. *Theriogenology* 66(2):283-291.

Santos R, Tharasanit T, Van Haeften T, Figueiredo J, Silva J, Van den Hurk R. 2007. Vitrification of goat preantral follicles enclosed in ovarian tissue by using conventional and solid-surface vitrification methods. *Cell and Tissue Research* 327(1):167-176.

Santos RMd, Barreta MH, Frajblat M, Cucco DC, Mezzalira JC, Bunn S, Cruz FB, Vieira AD, Mezzalira A. 2006. Vacuum-cooled liquid nitrogen increases the developmental ability of vitrified-warmed bovine oocytes. *Ciencia Rural* 36(5):1501-1506.

Saragusty J, Gacitua H, King R, Arav A. 2006. Post-mortem semen cryopreservation and characterization in two different endangered gazelle species (*Gazella gazella* and *Gazella dorcas*) and one subspecies (*Gazella gazelle acaiae*). *Theriogenology* 66(4):775-784.

Saragusty J, Gacitua H, Pettit MT, Arav A. 2007. Directional freezing of equine semen in large volumes. *Reproduction in Domestic Animals* 42(6):610-615.

Saragusty J, Gacitua H, Rozenboim I, Arav A. 2009a. Do physical forces contribute to cryodamage? *Biotechnology and Bioengineering* 104(4):719-728.

Saragusty J, Gacitua H, Rozenboim I, Arav A. 2009b. Protective effects of iodixanol during bovine sperm cryopreservation. *Theriogenology* 71(9):1425-1432.

Saragusty J, Gacitua H, Zeron Y, Rozenboim I, Arav A. 2009c. Double freezing of bovine semen. *Animal Reproduction Science* 115(1-4):10-17.

Saragusty J, Hermes R, Göritz F, Schmitt DL, Hildebrandt TB. 2009d. Skewed birth sex ratio and premature mortality in elephants. *Animal Reproduction Science* 115(1-4):247-254.

Saragusty J, Hildebrandt TB, Behr B, Knieriem A, Kruse J, Hermes R. 2009e. Successful cryopreservation of Asian elephant (*Elephas maximus*) spermatozoa. *Animal Reproduction Science* 115(1-4):255-266.

Saragusty J, Hildebrandt TB, Bouts T, Göritz F, Hermes R. 2010a. Collection and preservation of pygmy hippopotamus (*Choeropsis liberiensis*) semen. *Theriogenology* 74(4):652-657.

Saragusty J, Hildebrandt TB, Natan Y, Hermes R, Yavin S, Göritz F, Arav A. 2005. Effect of egg-phosphatidylcholine on the chilling sensitivity and lipid phase transition of Asian elephant (*Elephas maximus*) spermatozoa. *Zoo Biology* 24(3):233-245.

Saragusty J, Walzer C, Petit T, Stalder G, Horowitz I, Hermes R. 2010b. Cooling and freezing of epididymal sperm in the common hippopotamus (*Hippopotamus amphibius*). *Theriogenology* 74(7):1256-1263.

Saroff J, Mixner JP. 1955. The relationship of egg yolk and glycerol content of diluters and glycerol equilibration time to survival of bull spermatozoa after low temperature freezing. *Journal of Dairy Science* 38(3):292-297.

Sato T, Katagiri K, Gohbara A, Inoue K, Ogonuki N, Ogura A, Kubota A, Ogawa T. 2011. *In vitro* production of functional sperm in cultured neonatal mouse testes. *Nature* 471(7339):504-507.

Schenk JL, DeGrofft DL. 2003. Insemination of cow elk with sexed frozen semen. *Theriogenology* 59(1):514 (abstract).

Schiewe MC. 1991a. The science and significance of embryo cryopreservation. *Journal of Zoo and Wildlife Medicine* 22(1):6-22.

Schiewe MC, Bush M, Phillips LG, Citino S, Wildt DE. 1991b. Comparative aspects of estrus synchronization, ovulation induction, and embryo cryopreservation in the Scimitar-horned Oryx, Bongo, Eland, and Greater Kudu. *Journal of Experimental Zoology* 258:75-88.

Schlatt S, Foppiani L, Rolf C, Weinbauer GF, Nieschlag E. 2002a. Germ cell transplantation into X-irradiated monkey testes. *Human Reproduction* 17(1):55-62.

Schlatt S, Kim SS, Gosden R. 2002b. Spermatogenesis and steroidogenesis in mouse, hamster and monkey testicular tissue after cryopreservation and heterotopic grafting to castrated hosts. *Reproduction* 124(3):339-346.

Schmitt DL, Hildebrandt TB. 1998. Manual collection and characterization of semen from Asian elephants (*Elephas maximus*). *Animal Reproduction Science* 53(1-4):309-314.

Schmitt DL, Hildebrandt TB. 2000. Corrigendum to "Manual collection and characterization of semen from asian elephants" [Anim. Reprod. Sci. 53 (1998) 309-314]. *Animal Reproduction Science* 59(1-2):119.

Schneiders A, Sonksen J, Hodges JK. 2004. Penile vibratory stimulation in the marmoset monkey: a practical alternative to electro-ejaculation, yielding ejaculates of enhanced quality. *Journal of Medical Primatology* 33(2):98-104.

Schoysman R, Vanderzwalmen P, Nijs M, Segal L, Segal-Bertin G, Geerts L, van Roosendaal E, Schoysman D. 1993. Pregnancy after fertilisation with human testicular spermatozoa. *The Lancet* 342(8881):1237.

Schuster TG, Keller LM, Dunn RL, Ohl DA, Smith GD. 2003. Ultra-rapid freezing of very low numbers of sperm using cryoloops. *Human Reproduction* 18(4):788-795.

Schwarzenberger F, Francke R, Goltenboth R. 1993. Concentrations of faecal immunoreactive progestagen metabolites during the oestrous cycle and pregnancy in the black rhinoceros (*Diceros bicornis michaeli*). *Journal of Reproduction and Fertility* 98(1):285-291.

Schwarzenberger F, Mostl F, Palme R, Bamberg E. 1996. Fecal steroid analysis for non-invasive monitoring of reproductive status in farm, wild and zoo animals. *Animal Reproduction Science* 42:515-526.

Seki S, Mazur P. 2009. The dominance of warming rate over cooling rate in the survival of mouse oocytes subjected to a vitrification procedure. *Cryobiology* 59(1):75-82.

Sereni E, Bonu MA, Fava L, Sciajno R, Serrao L, Preti S, Distratis V, Borini A. 2008. Freezing spermatozoa obtained by testicular fine needle aspiration: a new technique. *Reproductive Biomedicine Online* 16(1):89-95.

Shaw J, Temple-Smith P, Trounson A, Lamden K. 1996. Fresh and frozen marsupial (*Sminthopsis crassicaudata*) ovaries develop after grafting to SCID mice. *Cryobiology* 33(6):631 (abstract).

Shefi S, Raviv G, Eisenberg ML, Weissenberg R, Jalalian L, Levron J, Band G, Turek PJ, Madgar I. 2006. Posthumous sperm retrieval: analysis of time interval to harvest sperm. *Human Reproduction* 21(11):2890-2893.

Sherman JK. 1954. Freezing and freeze-drying of human spermatozoa. *Fertility and Sterility* 5(4):357-371.

Sherman JK. 1957. Freezing and freeze-drying of bull spermatozoa. *American Journal of Physiology* 190(2):281-286.

Sherman JK. 1963. Improved methods of preservation of human spermatozoa by freezing and freeze-drying. *Fertility and Sterility* 14:49-64.

Shi D, Lu F, Wei Y, Cui K, Yang S, Wei J, Liu Q. 2007. Buffalos (Bubalus bubalis) cloned by nuclear transfer of somatic cells. *Biology of Reproduction* 77(2):285-291.

Shiels PG, Kind AJ, Campbell KHS, Wilmut I, Waddington D, Colman A, Schnieke AE. 1999. Analysis of telomere length in Dolly, a sheep derived by nuclear transfer. *Cloning* 1(2):119-125.

Shimshony A. 1988. Foot and mouth disease in the Mountain gazelle in Israel. *Revue Scientifique et Technique* 7(4):917-923.

Shimshony A, Orgad U, Baharav D, Prudovsky S, Yakobson B, Bar Moshe B, Dagan D. 1986. Malignant foot-and-mouth disease in Mountain gazelles. *Veterinary Record* 119(8):175-176.

Shin T, Kraemer D, Pryor J, Liu L, Rugila J, Howe L, Buck S, Murphy K, Lyons L, Westhusin M. 2002. A cat cloned by nuclear transplantation. *Nature* 415(6874):859.

Shinohara T, Inoue K, Ogonuki N, Kanatsu-Shinohara M, Miki H, Nakata K, Kurome M, Nagashima H, Toyokuni S, Kogishi K and others. 2002. Birth of offspring following transplantation of cryopreserved immature testicular pieces and *in-vitro* microinsemination. *Human Reproduction* 17(12):3039-3045.

Shinohara T, Kato M, Takehashi M, Lee J, Chuma S, Nakatsuji N, Kanatsu-Shinohara M, Hirabayashi M. 2006. Rats produced by interspecies spermatogonial transplantation in mice and *in vitro* microinsemination. *Proceedings of the National Academy of Sciences of the United States of America* 103(37):13624-13628.

Si W, Hildebrandt TB, Reid C, Krieg R, Ji W, Fassbender M, Hermes R. 2006. The successful double cryopreservation of rabbit (*Oryctolagus cuniculus*) semen in large volume using the directional freezing technique with reduced concentration of cryoprotectant. *Theriogenology* 65(4):788-798.

Si W, Lu Y, He X, Ji S, Niu Y, Tan T, Ji W. 2009. Improved survival by cryopreservation rhesus macaque (*Mcaca mulatta*) spermatozoa with directional freezing technique. *Reproduction, Fertility and Development* 22(1):217 (abstract).

Si W, Lu Y, He X, Ji S, Niu Y, Tan T, Ji W. 2010. Directional freezing as an alternative method for cryopreserving rhesus macaque (*Macaca mulatta*) sperm. *Theriogenology* 74(8):1431-1438.

Silakes S, Bart AN. 2010. Ultrasound enhanced permeation of methanol into zebrafish, Danio rerio, embryos. *Aquaculture* 303(1-4):71-76.

Silber SJ, Pineda JA, DeRosa M, Gorman KS, Patrizio P, Gosden RG. 2007. Comparison of ovarian cortical tissue grafting vs. intact whole ovary microvascular homotransplantation and allotransplantation for patients with premature ovarian failure. *Fertility and Sterility* 88(Supplement 1):S45 (abstract).

Skidmore JA, Loskutoff NM. 1999. Developmental competence *in vitro* and *in vivo* of cryopreserved expanding blastocysts from the dromedary camel (*Camelus dromedarius*). *Theriogenology* 51(1):293 (abstract).

Slade NP, Takeda T, Squires EL, Elsden RP, Seidel GE, Jr. 1985. A new procedure for the cryopreservation of equine embryos. *Theriogenology* 24(1):45-58.

Smith AU, Parkes AS. 1951. Preservation of ovarian tissue at low temperatures. *The Lancet* 258(6683):570-572.

Snedaker AK, Honaramooz A, Dobrinski I. 2004. A game of cat and mouse: Xenografting of testis tissue from domestic kittens results in complete cat spermatogenesis in a mouse host. *Journal of Andrology* 25(6):926-930.

Snow M, Cox SL, Jenkin G, Trounson A, Shaw J. 2002. Generation of live young from xenografted mouse ovaries. *Science* 297(5590):2227.

Soler JP, Mucci N, Kaiser GG, Aller J, Hunter JW, Dixon TE, Alberio RH. 2007. Multiple ovulation and embryo transfer with fresh, frozen and vitrified red deer (*Cervus elaphus*) embryos in Argentina. *Animal Reproduction Science* 102(3-4):322-327.

Soulé ME. 1991. Conservation: Tactics for a constant crisis. *Science* 253(5021):744-750.

Spallanzani L. 1776. Opuscoli di fisica animale e vigitabile. Opuscolo II. Observazioni e sperienze intorno ai vermicelli spermaici dell' homo e degli animali. Modena: Presso la Societa Tipografica.

Spindler RE, Pukazhenthi BS, Wildt DE. 2000. Oocyte metabolism predicts the development of cat embryos to blastocyst *in vitro*. *Molecular Reproduction and Development* 56(2):163-171.

Steponkus PL, Myers SP, Lynch DV, Gardner L, Bronshteyn V, Leibo SP, Rall WF, Pitt RE, Lin TT, MacIntyre RJ. 1990. Cryopreservation of *Drosophila melanogaster* embryos. *Nature* 345(6271):170-172.

Stover J, Evans J. Interspecies embryos transfer from gaur (*Bos gaurus*) to domestic Holstein cattle (*Bos taurus*) at the New York Zoological Park. In: Stover J, Evans J, editors; 1984 June 10-14, 1984; University of Illinois at Urbana-Champaign, Illinois, USA. p 243.1-243.3 (abstract).

Strzezek J, Fraser L, Kuklinska M, Dziekonska A, Lecewicz M. 2004. Effects of dietary supplementation with polyunsaturated fatty acids and antioxidants on biochemical characteristics of boar semen. *Reproductive Biology* 4(3):271-287.

Stukenborg J-B, Schlatt S, Simoni M, Yeung C-H, Elhija MA, Luetjens CM, Huleihel M, Wistuba J. 2009. New horizons for *in vitro* spermatogenesis? An update on novel three-dimensional culture systems as tools for meiotic and post-meiotic differentiation of testicular germ cells. *Molecular Human Reproduction* 15(9):521-529.

Sugiyama R, Nakagawa K, Shirai A, Sugiyama R, Nishi Y, Kuribayashi Y, Inoue M. 2010. Clinical outcomes resulting from the transfer of vitrified human embryos using a new device for cryopreservation (plastic blade). *Journal of Assisted Reproduction and Genetics* 27(4):161-167.

Summers PM, Shephard AM, Taylor CT, Hearn JP. 1987. The effects of cryopreservation and transfer on embryonic development in the common marmoset monkey, *Callithrix jacchus*. *Journal of Reproduction and Fertility* 79(1):241-250.

Sun WQ, Leopold AC, Crowe LM, Crowe JH. 1996. Stability of dry liposomes in sugar glasses. *Biophysical Journal* 70(4):1769-1776.

Sun X, Li Z, Yi Y, Chen J, Leno GH, Engelhardt JF. 2008. Efficient term development of vitrified ferret embryos using a novel pipette chamber technique. *Biology of Reproduction* 79(5):832-840.

Suzuki H, Asano T, Suwa Y, Abe Y. 2009. Successful delivery of pups from cryopreserved canine embryos. *Biology of Reproduction* 81(1 supplement):619 (abstract).

Swain JE, Miller RR, Jr. 2000. A postcryogenic comparison of membrane fatty acids of elephant spermatozoa. *Zoo Biology* 19:461-473.

Swanson WF. 2001. Reproductive biotechnology and conservation of the forgotten felids - the small cats; 17-18 Januar, 2001; Omaha. p 100-120.

Swanson WF. 2003. Research in nondomestic species: Experiences in reproductive physiology research for conservation of endangered felids. *ILAR Journal* 44(4):307-316.

Swanson WF. 2006. Application of assisted reproduction for population management in felids: The potential and reality for conservation of small cats. *Theriogenology* 66(1):49-58.

Swanson WF, Brown JL. 2004. International training programs in reproductive sciences for conservation of Latin American felids. *Animal Reproduction Science* 82-83:21-34.

Swanson WF, Paz RCR, Morais RN, Gomes MLF, Moraes W, Adania CH. 2002. Influence of species and diet on efficiency of *in vitro* fertilization in two endangered Brazilian felids - the ocelot (*Leopardus pardalis*) and tigrina (*Leopardus tigrinus*). *Theriogenology* 57(1):593 (abstract).

Takagi M, Kim IH, Izadyar F, Hyttel P, Bevers MM, Dieleman SJ, Hendriksen PJ, Vos PL. 2001. Impaired final follicular maturation in heifers after superovulation with recombinant human FSH. *Reproduction* 121(6):941-951.

Tedder RS, Zuckerman MA, Goldstone AH, Hawkins AE, Fielding A, Briggs EM, Irwin D, Blair S, Gorman AM, Patterson KG and others. 1995.

Hepatitis B transmission from contaminated cryopreservation tank. *Lancet* 346(8968):137-140.

Thibier M. 2006. Data Retrieval Committee Annual Report: Transfers of both *in vivo* derived and *in vitro* produced embryoa in cattle still on the rise and contrasted trends in other species in 2005. *International Embryo Transfer Society Newsletter* 24(4):12-18.

Thibier M. 2009. Data Retrieval Committee statistics of embryo transfer - year 2008. The worldwide statistics of embryo transfers in farm animals. *International Embryo Transfer Society Newsletter* 27(4):13-19.

Thomas CD, Cameron A, Green RE, Bakkenes M, Beaumont LJ, Collingham YC, Erasmus BFN, de Siqueira MF, Grainger A, Hannah L and others. 2004. Extinction risk from climate change. *Nature* 427(6970):145-148.

Thomson JA, Itskovitz-Eldor J, Shapiro SS, Waknitz MA, Swiergiel JJ, Marshall VS, Jones JM. 1998. Embryonic stem cell lines derived from human blastocysts. *Science* 282(5391):1145-1147.

Thomson LK, Fleming SD, Barone K, Zieschang J-A, Clark AM. 2010. The effect of repeated freezing and thawing on human sperm DNA fragmentation. *Fertility and Sterility* 93(4):1147-1156.

Thundathil J, Whiteside D, Shea B, Ludbrook D, Elkin B, Nishi J. 2007. Preliminary assessment of reproductive technologies in wood bison (*Bison bison athabascae*): Implications for preserving genetic diversity. *Theriogenology* 68(1):93-99.

Toyooka Y, Tsunekawa N, Akasu R, Noce T. 2003. Embryonic stem cells can form germ cells *in vitro*. *Proceedings of the National Academy of Science of the Uunited States of America* 100(20):11457-11462.

Trimeche A, Renard P, Tainturier D. 1998. A procedure for Poitou jackass sperm cryopreservation. *Theriogenology* 50(5):793-806.

Trounson A, Mohr L. 1983. Human pregnancy following cryopreservation, thawing and transfer of an eight-cell embryo. *Nature* 305(5936):707-709.

Tsang WH, Chow KL. 2009. Mouse embryo cryopreservation utilizing a novel high-capacity vitrification spatula. *BioTechniques* 46(7):550-552.

Underwood SL, Bathgate R, Ebsworth M, Maxwell WMC, Evans G. 2010. Pregnancy loss in heifers after artificial insemination with frozen-thawed, sex-sorted, re-frozen-thawed dairy bull sperm. *Animal Reproduction Science* 118(1):7-12.

Underwood SL, Bathgate R, Maxwell WMC, O'Donnell M, Evans G. 2007. Pregnancies after artificial insemination of frozen-thawed, sex-sorted, re-frozen-thawed dairy bull sperm. *Reproduction in Domestic Animals* 42(S2):78 (abstract).

Uteshev VK, Mel'nikova EV, Kaurova SA, Nikitin VA, Gakhova EN, Karnaukhov VN. 2002. Fluorescence analysis of cryopreserved totipotent cells of amphibian embryos. *Biophysics* 47(3):506-512.

Vajta G, Holm P, Greve T, Callesen H. 1997. Vitrification of porcine embryos using the Open Pulled Straw (OPS) method. *Acta Veterinaria Scandinavica* 38(4):349-352.

Vajta G, Holm P, Kuwayama M, Booth PJ, Jacobsen H, Greve T, Callesen H. 1998. Open Pulled Straw (OPS) vitrification: a new way to reduce cryoinjuries of bovine ova and embryos. *Molecular Reproduction and Development* 51(1):53-58.

Van Thuan N, Wakayama S, Kishigami S, Wakayama T. 2005. New preservation method for mouse spermatozoa without freezing. *Biology of Reproduction* 72(2):444-450.

Vanderzwalmen P, Bertin G, Debauche C, Standaart V, Schoysman E. 2000. "*In vitro*" survival of metaphase II oocytes (MII) and blastocysts after vitrification in a hemi-straw (HS) system. *Fertility and Sterility* 74(3, Supplement 1):S215-S216 (abstract).

Vendramini OM, Bruyas JF, Fieni F, Battut I, Tainturier D. 1997. Embryo transfer in Poitou donkeys, preliminary results. *Theriogenology* 47(1):409 (abstract).

Veprintsev BN, Rott NN. 1979. Conserving genetic resources of animal species. *Nature* 280(5724):633-634.

Verza S, Jr., Feijo CM, Esteves SC. 2009. Resistance of human spermatozoa to cryoinjury in repeated cycles of thaw-refreezing. *International Brazilian Journal of Urology* 35(5):581-591.

Vidament M, Vincent P, Yvon JM, Bruneau B, Martin FX. 2005. Glycerol in semen extender is a limiting factor in the fertility in asine and equine species. *Animal Reproduction Science* 89(1-4):302-305 (abstract).

von Baer A, Del Campo MR, Donoso X, Toro F, von Baer L, Montecinos S, Rodriguez-Martinez H, Palasz T. 2002. Vitrification and cold storage of llama (*Lama glama*) hatched blastocysts. *Theriogenology* 57(1):489 (abstract).

Wakayama T, Yanagimachi R. 1998. Development of normal mice from oocytes injected with freeze-dried spermatozoa. *Nature Biotechnology* 16(7):639-641.

Wake DB, Vredenburg VT. 2008. Colloquium paper: are we in the midst of the sixth mass extinction? A view from the world of amphibians. *Proceedings of the National Academy of Science of the Uunited States of America* 105 Suppl 1:11466-11473.

Walmsley R, Cohen J, Ferrara-Congedo T, Reing A, Garrisi J. 1998. The first births and ongoing pregnancies associated with sperm cryopreservation within evacuated egg zonae. *Human Reproduction* 13(suppl 4):61-70.

Walpole M, Almond REA, Besancon C, Butchart SHM, Campbell-Lendrum D, Carr GM, Collen B, Collette L, Davidson NC, Dulloo E and others. 2009. Tracking progress toward the 2010 biodiversity target and beyond. *Science* 325(5947):1503-1504.

Wang W. 2000. Lyophilization and development of solid protein pharmaceuticals. *International Journal of Pharmaceutics* 203(1-2):1-60.

Wang X, Catt S, Pangestu M, Temple-Smith P. 2011. Successful *in vitro* culture of pre-antral follicles derived from vitrified murine ovarian tissue: oocyte maturation, fertilization, and live births. *Reproduction* 141(2):183-191.

Wang X, Chen H, Yin H, Kim SS, Lin Tan S, Gosden RG. 2002. Fertility after intact ovary transplantation. *Nature* 415(6870):385.

Wang Y, Xiao Z, Li L, Fan W, Li S-W. 2008. Novel needle immersed vitrification: a practical and convenient method with potential advantages in mouse and human ovarian tissue cryopreservation. *Human Reproduction* 23(10):2256-2265.

Ward MA, Kaneko T, Kusakabe H, Biggers JD, Whittingham DG, Yanagimachi R. 2003. Long-term preservation of mouse spermatozoa after freeze-drying and freezing without cryoprotection. *Biology of Reproduction* 69(6):2100-2108.

Watson PF. 2000. The causes of reduced fertility with cryopreserved semen. *Animal Reproduction Science* 60-61:481-492.

Weiss L, Armstrong JA. 1960. Structural changes in mammalian cells associated with cooling to -79°C. *The Journal of Biophysical and Biochemical Cytology* 7(4):673-678.

Weissman A, Gotlieb L, Colgan T, Jurisicova A, Greenblatt EM, Casper RF. 1999. Preliminary experience with subcutaneous human ovarian cortex transplantation in the NOD-SCID mouse. *Biology of Reproduction* 60(6):1462-1467.

Westh P, Ramløv H. 1991. Trehalose accumulation in the tardigrade *Adorybiotus coronifer* during anhydrobiosis. *Journal of Experimental Zoology* 258(3):303-311.

Whittingham DG. 1971. Survival of mouse embryos after freezing and thawing. *Nature* 233(5315):125-126.

Whittingham DG. 1975. Survival of rat embryos after freezing and thawing. *Journal of Reproduction and Fertility* 43(3):575-578.

Whittingham DG. 1977. Fertilization *in vitro* and development to term of unfertilized mouse oocytes previously stored at -196°C. *Journal of Reproduction and Fertility* 49(1):89-94.

Whittingham DG, Adams CE. 1974. Low temperature preservation of rabbit embryos. *Cryobiology* 11(6):560-561 (abstract).

Whittingham DG, Adams CE. 1976. Low temperature preservation of rabbit embryos. *Journal of Reproduction and Fertility* 47(2):269-274.

Whittingham DG, Leibo SP, Mazur P. 1972. Survival of mouse embryos frozen to -196° and -269°C. *Science* 178(4059):411-414.

Wiese RJ. 2000. Asian elephants are not self-sustaining in North America. *Zoo Biology* 19(5):299-309.

Wildt D, Pukazhenthi B, Brown J, Monfort S, Howard J, Roth T. 1995. Spermatology for understanding, managing and conserving rare species. *Reproduction, Fertility and Development* 7(4):811-824.

Wildt DE. 1992. Genetic resource banks for conserving wildlife species: justification, examples and becoming organized on a global basis. *Animal Reproduction Science* 28(1-4):247-257.

Wildt DE, Rall WF, Critser JK, Monfort SL, Seal US. 1997. Genome resource banks: living collections for biodiversity conservation. *Bioscience* 47(10):689-698.

Wildt DE, Schiewe MC, Schmidt PM, Goodrowe KL, Howard JG, Phillips LG, O'Brien SJ, Bush M. 1986. Developing animal model systems for embryo technologies in rare and endangered wildlife. *Theriogenology* 25(1):33-51.

Wildt DE, Wemmer C. 1999. Sex and wildlife: the role of reproductive science in conservation. *Biodiversity and Conservation* 8(7):965-976.

Willadsen S, Polge C, Rowson LEA. 1978. The viability of deep-frozen cow embryos. *Journal of Reproduction and Fertility* 52(2):391-393.

Willadsen SM, Polge C, Rowson LEA, Moor RM. 1974. Preservation of sheep embryos in liquid nitrogen. *Cryobiology* 11(6):560 (abstract).

Willadsen SM, Polge C, Rowson LEA, Moor RM. 1976. Deep freezing of sheep embryos. *Journal of Reproduction and Fertility* 46(1):151-154.

Wilmut I. 1972. The effect of cooling rate, warming rate, cryoprotective agent and stage of development of survival of mouse embryos during freezing and thawing. *Life Sciences* 11(22, Part 2):1071-1079.

Wilmut I, Rowson LE. 1973. Experiments on the low-temperature preservation of cow embryos. *The Veterinary Record* 92(26):686-690.

Wilmut I, Schnieke AE, McWhir J, Kind AJ, Campbell KH. 1997. Viable offspring derived from fetal and adult mammalian cells. *Nature* 385(6619):810-813.

Wolf DP, Vandevoort CA, Meyer-Haas GR, Zelinski-Wooten MB, Hess DL, Baughman WL, Stouffer RL. 1989. *In vitro* fertilization and embryo transfer in the rhesus monkey. *Biology of Reproduction* 41(2):335-346.

Wolfe BA, Wildt DE. 1996. Development to blastocysts of domestic cat oocytes matured and fertilized *in vitro* after prolonged cold storage. *Journal of Reproduction and Fertility* 106(1):135-141.

Wolvekamp MC, Cleary ML, Cox SL, Shaw JM, Jenkin G, Trounson AO. 2001. Follicular development in cryopreserved common Wombat ovarian tissue xenografted to nude rats. *Animal Reproduction Science* 65(1-2):135-147.

Womersley C, Ching C. 1989. Natural dehydration regimes as a prerequisite for the successful induction of anhydrobiosis in the nematode *Rotylenchulus reniformis*. *Journal of Experimental Biology* 143(1):359-372.

Wood TC, Wildt DE. 1997. Effect of the quality of the cumulus-oocyte complex in the domestic cat on the ability of oocytes to mature, fertilize and develop into blastocysts *in vitro*. *Journal of Reproduction and Fertility* 110(2):355-360.

Woods EJ, Newton L, Critser JK. 2010. A novel closed system vial with sentinel test segment for sperm cryopreservation. *Fertility and Sterility* 94(4, Supplement 1):S239-S239.

Woods GL, White KL, Vanderwall DK, Li GP, Aston KI, Bunch TD, Meerdo LN, Pate BJ. 2003. A mule cloned from fetal cells by nuclear transfer. *Science* 301(5636):1063.

Yamamoto Y, Oguri N, Tsutsumi Y, Hachinohe Y. 1982. Experiments in the freezing and storage of equine embryos. *Journal of Reproduction and Fertility Supplement* 32:399-403.

Yavin S, Arav A. 2001. Development of immature bovine oocytes vitrified by minimum drop size technique and a new vitrification apparatus (VIT-MASTER). *Cryobiology* 43(4):331 (abstract).

Yavin S, Arav A. 2007. Measurement of essential physical properties of vitrification solutions. *Theriogenology* 67(1):81-89.

Yavin S, Aroyo A, Roth Z, Arav A. 2009. Embryo cryopreservation in the presence of low concentration of vitrification solution with sealed pulled straws in liquid nitrogen slush. *Human Reproduction* 24(4):797-804.

Yeoman RR, Gerami-Naini B, Mitalipov S, Nusser KD, Widmann-Browning AA, Wolf DP. 2001. Cryoloop vitrification yields superior survival of Rhesus monkey blastocysts. *Human Reproduction* 16(9):1965-1969.

Yeoman RR, Wolf DP, Lee DM. 2005. Coculture of monkey ovarian tissue increases survival after vitrification and slow-rate freezing. *Fertility and Sterility* 83(4, Supplement 1):1248-1254.

Yin H, Wang X, Kim SS, Chen H, Tan SL, Gosden RG. 2003. Transplantation of intact rat gonads using vascular anastomosis: effects of cryopreservation, ischaemia and genotype. *Human Reproduction* 18(6):1165-1172.

Yoon TK, Lee DR, Cha SK, Chung HM, Lee WS, Cha KY. 2007. Survival rate of human oocytes and pregnancy outcome after vitrification using slush nitrogen in assisted reproductive technologies. *Fertility and Sterility* 88(4):952-956.

Yu I, Leibo SP. 2002. Recovery of motile, membrane-intact spermatozoa from canine epididymides stored for 8 days at 4°C. *Theriogenology* 57(3):1179-1190.

Yushchenko NP. 1957. Proof of the possibility of preserving mammalian spermatozoa in a dried state. *Proceedings of the Lenin Academy of Agricultural Sciences of the USSR* 22:37-40.

Zambelli D, Cunto M. 2006. Semen collection in cats: Techniques and analysis. *Theriogenology* 66(2):159-165.

Zambelli D, Prati F, Cunto M, Iacono E, Merlo B. 2008. Quality and *in vitro* fertilizing ability of cryopreserved cat spermatozoa obtained by urethral catheterization after medetomidine administration. *Theriogenology* 69(4):485-490.

Zeilmaker GH, Alberda AT, van Gent I, Rijkmans CM, Drogendijk AC. 1984. Two pregnancies following transfer of intact frozen-thawed embryos. *Fertility and Sterility* 42(2):293-296.

Zeng W, Avelar GF, Rathi R, Franca LR, Dobrinski I. 2006. The length of the spermatogenic cycle is conserved in porcine and ovine testis xenografts. *Journal of Andrology* 27(4):527-33.

Zeron Y, Pearl M, Borochov A, Arav A. 1999. Kinetic and temporal factors influence chilling injury to germinal vesicle and mature bovine oocytes. *Cryobiology* 38(1):35-42.

Zeron Y, Sklan D, Arav A. 2002a. Effect of polyunsaturated fatty acid supplementation on biophysical parameters and chilling sensitivity of ewe oocytes. *Molecular Reproduction and Development* 61(2):271-278.

Zeron Y, Tomczak M, Crowe J, Arav A. 2002b. The effect of liposomes on thermotropic membrane phase transitions of bovine spermatozoa and oocytes: implications for reducing chilling sensitivity. *Cryobiology* 45(2):143-152.

Zhang T, Rawson DM. 1995. Studies on chilling sensitivity of zebrafish (*Brachydanio rerio*) embryos. *Cryobiology* 32(3):239-246.

Zhang T, Rawson DM. 1996. Feasibility studies on vitrification of intact zebrafish (*Brachydanio rerio*) embryos. *Cryobiology* 33(1):1-13.

INDEX